Alteration of Native Hawaiian Vegetation

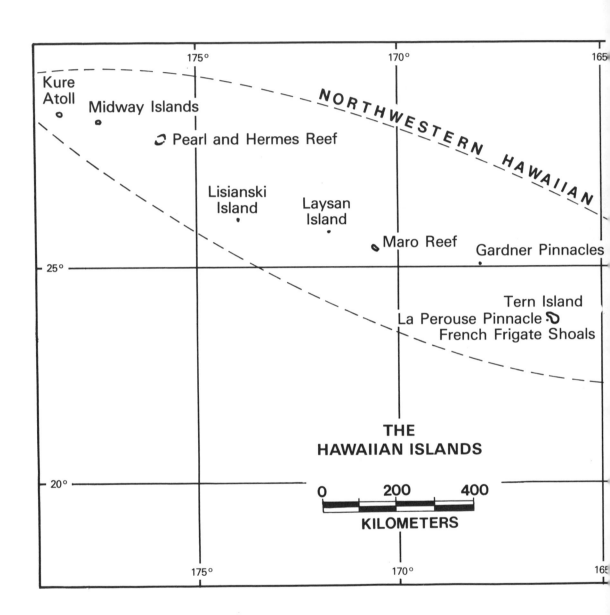

Kure
Atoll

Midway Islands

Pearl and Hermes Reef

NORTHWESTERN HAWAIIAN

Lisianski
Island

Laysan
Island

Maro Reef   Gardner Pinnacles

25°

Tern Island
La Perouse Pinnacle
French Frigate Shoals

175°   170°   165°

THE
HAWAIIAN ISLANDS

0   200   400

KILOMETERS

20°

175°   170°   165°

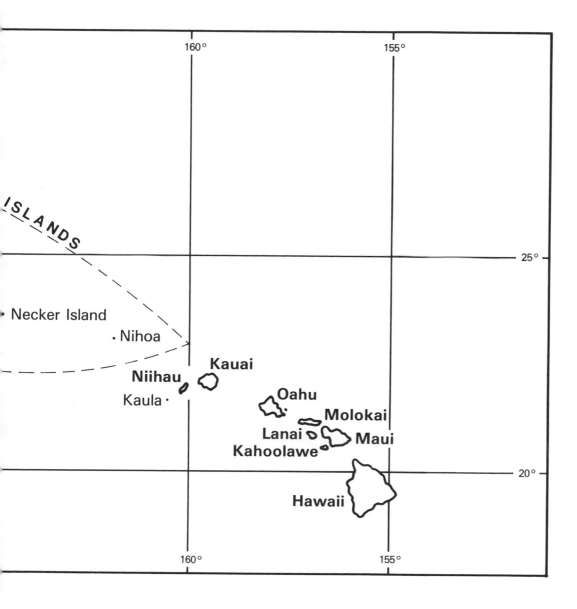

Alteration of Native Hawaiian Vegetation

Effects of Humans, Their Activities and Introductions

By

Linda W. Cuddihy and Charles P. Stone

University of Hawaii Cooperative National Park Resources Studies Unit
3190 Maile Way · Honolulu, Hawai'i

Support for publication was provided by
the National Park Service

First printing 1990
Second printing 1990
Third printing 1993

Library of Congress Cataloging-in-Publication Data

Cuddihy, Linda W.
   Alteration of native Hawaiian vegetation : effects of humans,
their activities and introductions / by Linda W. Cuddihy and Charles
P. Stone.
      p.   cm.
   Includes bibliographical references.
   ISBN 0-8248-1308-1 : $18.00
   1. Botany--Hawaii.   2. Man--Influence on nature--Hawaii.
3. Botany--Hawaii--Ecology.   I. Stone, Charles P.   II. Title.
QK473.H4C83   1990
581.5'24'09969--dc20   89-20592
                                                                CIP

Distributed by
University of Hawaii Press
Honolulu, Hawaii 96822

Contents

Authors' Preface

This book began as a limited review of what is known about the decline of native vegetation in Hawai'i since the arrival of humans. The purpose of documenting vegetation alteration was to eventually attempt to relate the changes to declines and extinctions of Hawai'i's native birds. Reviews of other likely long-term influences on birds, such as predation and avian diseases, competition, and declines in avian food resources were also planned. In addition, separate reviews of population information on each of Hawai'i's native avian species were begun, and these have been largely accomplished by Winston Banko under private contract with the Cooperative National Park Resources Studies Unit at the University of Hawaii at Manoa. The final objective of the entire project was to compare all possible causes (including vegetation changes) of the declines and extinctions of Hawai'i's birds -- a loss of at least 66% of the known historic and fossil species -- and thoroughly explore the relative importance of different factors. This objective has yet to be accomplished, although it is likely that it will eventually be done.

Because the story of the alteration of Hawai'i's vegetation by humans, their activities and introductions is of wide interest in Hawai'i; because our efforts to summarize and evaluate much unpublished and difficult-to-access material and bring together overviews on different subjects went beyond the initial intent; and because the eventual use of this review to evaluate native bird declines is some time away, we decided to publish the information on vegetation alteration for a wider audience than initially planned. The review is not exhaustive for each subject covered -- in many cases, examples were used to illustrate points -- but the literature citations are extensive and provide points of departure for further study.

Scientific names are italicized throughout the book, and only scientific names are used if common names do not exist. Scientific names are given with common names upon first usage in each section. Hawaiian words are italicized where first used in each section. Macrons and glottal stops, or 'okina, are used on all words of Hawaiian origin where appropriate, except some names of organizations or institutions. Spelling of Hawaiian words is according to Pukui and Elbert 1981 for the most part; some common Hawaiian names of plants were taken from Porter (1972). Hawaiian place names follow Pukui et al., 1986, except where prevailing common usage differs. Scientific names of flowering plants are according to Wagner et al. (in press), except for a few rare plants where older nomenclature (St. John 1973) is retained. Names of plants used by the authors we cited were changed to conform to Wagner et al. (in press). Fern nomenclature is from C.H. Lamoureux (unpubl. ms). Scientific names of animals are taken from the sources cited, with the exception of bird nomenclature, which is from Pratt et al. (1987).

Acknowledgements

We especially thank eight individuals for their critical reviews, additions, and useful suggestions on manuscript drafts: Stephen Anderson, Dorothy Barrère, James Jacobi, Lloyd Loope, Dieter Mueller-Dombois, Clifford Smith, Danielle Stone, and Timothy Tunison. Lloyd Loope also brought to our attention much pertinent literature, especially related to the island of Maui. Jim Jacobi and Steve Anderson shared with us their unpublished manuscripts and data and allowed us to refer to their efforts.

Acknowledgement is due the authors of papers that are "in press" in the volume entitled *Alien plant invasions in native ecosystems of Hawai'i: management and research.* Senior or sole authors are: Steve Anderson, Derral Herbst, Jim Jacobi, James Juvik, Anne Marie LaRosa, Lloyd Loope, George Markin, Larry Nakahara, Clifford Smith, Tim Tunison, Peter Vitousek, Warren Wagner, Lyndon Wester, and Rylan Yee. We appreciate their cooperation.

Warren Wagner, Derral Herbst, and Sy Sohmer provided us with a review draft of their landmark book, the *Manual of flowering plants of Hawai'i,* and allowed us to cite it frequently. David Chai, Laura Huenneke, Lloyd Loope, Art Medeiros, The Nature Conservancy of Hawaii, Tim Tunison, and Peter Vitousek also allowed us to use unpublished information.

We are indebted to the University of Hawaii Press staff, especially Jan Heavenridge and Lucy Aono, who advised us on production of the text and artwork and encouraged our efforts. We also thank Cliff Smith, Director of the University of Hawaii Cooperative National Park Resources Studies Unit, and Bruce M. Kilgore, National Park Service Chief of the Division of Natural Resources and Research (Western Region), for affording us the flexibility to complete the project. Cliff also facilitated arrangements with the University of Hawaii Press.

The research staff at Hawaii Volcanoes National Park helped us in numerous ways. Thanks to Joan Yoshioka for meticulous maps and cover art; to Steve Anderson and Michelle Fulton for helping us with computer expertise and proofreading; and finally, we thank Danielle Stone for her perseverance and accuracy in word processing, editing, proofreading, and preparation of the numerous drafts necessary for the camera-ready production of this book.

Mahalo to Samuel M. Gon III, who provided us with the computer technique necessary to print macrons in Hawaiian words.

This volume was partially funded by National Park Service Contract CX-8092-2840-182.

INTRODUCTION

The Hawaiian Archipelago, located between 18 and 22° North latitude in the central Pacific Ocean, is one of the most isolated island groups in the world. A distance of more than 3,200 km (2,000 mi) separates these Islands from any major land mass. The Archipelago, spanning 2,400 km (1,500 mi), is often considered in terms of the eight "main" islands (Hawai'i to Ni'ihau) and the Northwestern Hawaiian Islands (Nihoa to Kure Atoll).

The Islands are volcanic in origin and were formed as the Pacific Plate moved northwest over a fixed melting anomaly or "hot spot" (Wilson 1963). The youngest of the main islands is on the southeastern end of the chain, and the islands to the northwest are progressively older. The oldest of the main Hawaiian Islands is Kaua'i, parts of which are estimated to be 5.6 million years in age (MacDonald and Abbott 1979; Rotondo et al. 1981). The youngest and largest island in the group is volcanically active Hawai'i, parts of which are estimated at 0.1-0.5 million years in age (MacDonald and Abbott 1979). Two anomalies in this chain of progressively aged islands are Necker and Wentworth Seamount, which are found northwest of Kaua'i and are much older than their neighbors. These landforms apparently arose elsewhere in the Pacific (to the southeast) and later integrated with the Hawaiian Chain as the Pacific Plate moved over the "hot spot." Island integration is of interest from a biological standpoint as well as a geological one, since the flora of Necker and other islands may have contributed to the development of the Hawaiian flora by the mixing of their biota (Rotondo et al. 1981). (Geographic relationships of the eight main Islands and place names used in the text are given Figures 1-3. The entire Archipelago is shown in the frontispiece map.)

The Hawaiian flora is not derived from any one source but rather developed in isolation for millions of years. The affinities of the present flora are primarily with plants of the Indo-Pacific, Americas, and South Pacific (Austral), although pantropic and boreal elements are also present; the affinities of some Hawaiian plants are unknown (Fosberg 1948a). The Malaysian and Australian elements of the Hawaiian flora are particularly pronounced: 80 genera (40%) of native flowering plants in Hawai'i either occur also in Malaysia or are closely allied to Malaysian genera (Skottsberg 1941). It is estimated that 272 original immigrants could account for the existing highly endemic angiosperm flora (Fosberg 1948a).

The number of endemic genera in the Hawaiian flora is remarkably high and has been previously estimated as 20 (Skottsberg 1941) to 28 (Fosberg 1948a). In a recent revision of the Hawaiian angiosperm flora, 32 endemic genera are recognized (Wagner et al., in press). The flora of the Hawaiian Islands contains approximately 1,000 (Wagner et al., in press) to 1,400 native species of flowering plants (St. John 1973). Some authors speculate that the native flora was much larger in the past, containing perhaps 20,000-30,000 species (Degener and Degener 1974), although analysis of pollen preserved in bogs does not support this contention (Selling 1948). The percentage of endemic angiosperm species has been estimated at 89% (Wagner et al., in press) to 96% (St. John 1973), in either case one of the highest percentages of endemism in the world. The number of Hawaiian fern species and varieties is approximately 168, with an estimated endemicity of 65% (Fosberg 1948a).

1

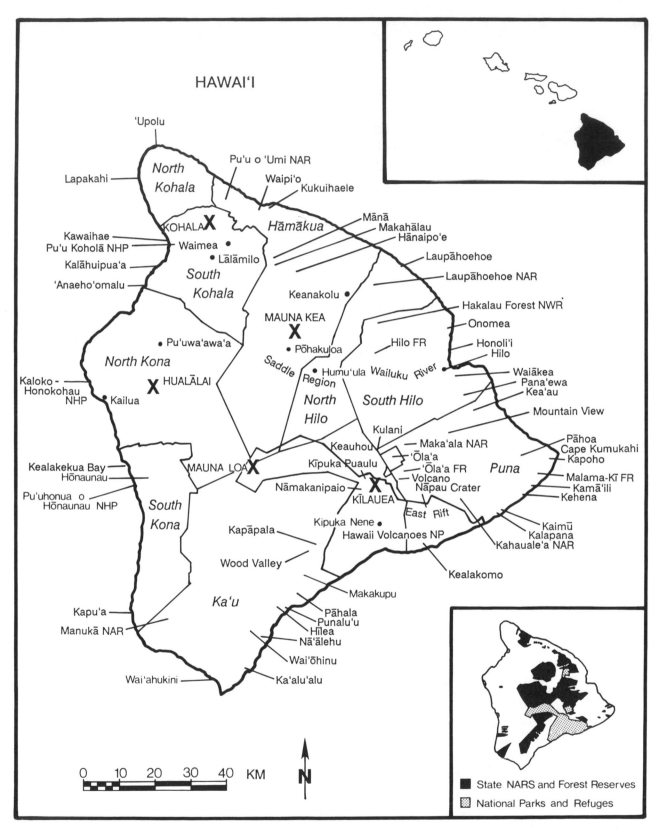

Figure 1. Location of places on Hawai‘i Island mentioned in text. (State and Federal lands are shown in lower inset, but National Historic Parks in western Hawai‘i are best seen on large map.)

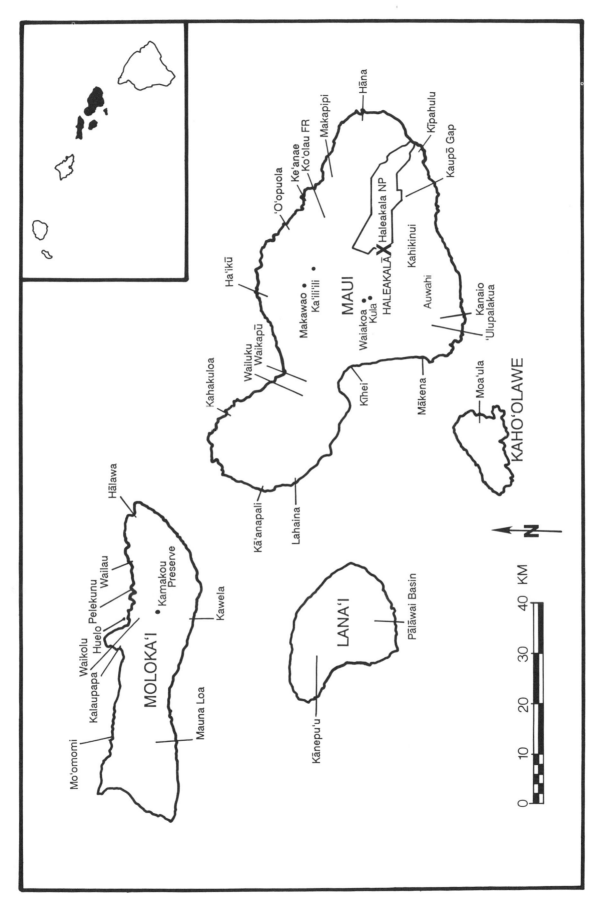

Figure 2. Location of places on Moloka'i, Lāna'i, Maui, and Kaho'olawe mentioned in text.

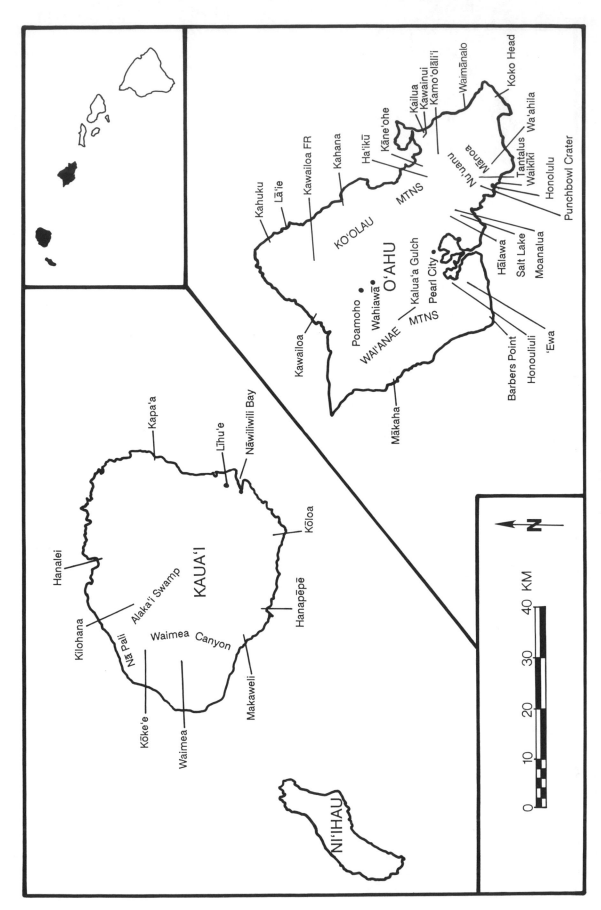

Figure 3. Location of places on Kaua'i and O'ahu mentioned in text.

VEGETATION OF THE HAWAIIAN ISLANDS

The vegetation of the Hawaiian Islands is relatively complex, often varying greatly over short distances. Recent classification attempts have recognized between 86 (Gagné and Cuddihy, in press) and 152 different plant communities (The Nature Conservancy of Hawaii 1987). This very large number of distinct communities in a small geographic area may be attributed to the great differences in rainfall, substrate, exposure, and topography that occur on each of the main Hawaiian Islands. In some cases, distinctions among communities may be partly an artifact of past disturbance, which has resulted in the disjunction and isolation of relictual stands of native vegetation formerly continuous over large areas.

Vegetation Classification Systems

Existing plant communities may be classified into broader vegetation zones based on one or more factors, such as elevation, moisture regime, substrate, topography, floristics, and plant physiognomy. Many such classification schemes have been developed to explain observed vegetation patterns in the Hawaiian Islands, the first by Gaudichaud-Beaupré in 1819 (St. John and Titcomb 1983; Jacobi, in press a). Since then, many botanists have produced new classification systems for Hawai'i or have refined earlier ones; 20 different Hawaiian vegetation classifications were compared and contrasted by Selling (1948). More recently, major systems were reviewed in depth by Jacobi (in press a).

Two important classification systems of the late 18th and early 19th century were those of Hillebrand (1888) and Rock (1913); both authors used their extensive field experience to classify observed vegetation into elevational zones. Rock also used climate and topography to characterize zones, for which he provided detailed floristic descriptions.

Perhaps the most widely used vegetation classification of the last several decades is that of Ripperton and Hosaka (1942), who produced a map of the main Hawaiian Islands with 10 zones based on elevation, topography, rainfall, and existing vegetation. Their zones were based on field sampling and emphasized differences in moisture regime (Jacobi, in press a). For more than 40 years, this classification of Hawaiian vegetation was generally adopted by most writers and was the basis of many subsequent modifications. A notable refinement of Ripperton and Hosaka's system was the biogeoclimatic zonation scheme developed by Krajina (1963), who incorporated edaphic factors and additional climate variables to distinguish 14 zones, each with a number of native or alien plant indicators.

Single-island classification systems have also been developed. The vegetation of O'ahu and Hawai'i has been the particular focus of several such efforts. Egler (1939) used moisture as a distinguishing environmental factor and recognized six vegetation zones on O'ahu. Egler did not base his classification on elevation or physiognomy, which was considered by Jacobi (in press a) to be a "significant departure from earlier classification systems." Hawai'i Island vegetation zones (or plant formations) were described and mapped by Robyns and Lamb (1939), who stressed the physiognomy of vegetation in their system (Jacobi, in press a). The zonation pattern of Robyns and Lamb appears very similar in outline to the Hawai'i map of Ripperton and Hosaka, but the former more thoroughly described existing natural vegetation.

Region-specific vegetation classifications were developed for Mauna Kea, Mauna Loa (Mueller-Dombois and Krajina 1968) and Hawaii Volcanoes National Park (Mueller-Dombois 1966). Both studies used extensive ground observations and photo-interpretation to produce profile diagrams detailing actual vegetation types in zones delimited by climate, elevation, and in some cases, substrate.

Most recently, Jacobi (in press b) developed a vegetation classification for the Hawaiian Islands in conjunction with a large-scale bird survey and mapping project (Scott et al. 1986). Emphasizing climate, with elevation secondary, Jacobi recognized 14 major vegetation (or "habitat") zones, each with one or more physiognomic subunits (e.g., grassland, shrubland, forest). The Nature Conservancy of Hawaii (1987) drew heavily on Jacobi's classification system in the organization of its natural community database, as did Gagné and Cuddihy (in press) in their synopsis of Hawaiian vegetation.

The following general discussion of natural vegetation of the Hawaiian Islands follows Jacobi's outline of vegetation zones and physiognomic units as modified by Gagné and Cuddihy and incorporates descriptions and locality information from both The Nature Conservancy of Hawaii (1987) and Gagné and Cuddihy (in press). Five elevational regions are recognized: coastal; lowland, with an upper limit between 500 m (1,640 ft) and 1,000 m (3,280 ft) elevation; montane (500-1,000 m to 2,000 m (6,560 ft) elevation); subalpine (> 2,000 m to 2,800 m (9,180 ft) elevation); and alpine (> 2,800 m). Within these, vegetation is categorized by moisture regime: dry, less than 1,250 mm (50 in.) annual precipitation; moist or mesic, 1,250-2,500 mm (50-100 in.); and wet, > 2,500 mm (100 in.) annual rainfall. Finally, five physiognomic types are used, based on the dominant plant life form: grassland, shrubland, forest, open forest, and parkland. Major vegetation zones of the windward sides of the Islands are presented first, starting at the coast and proceeding upslope to the summits; then lowland and montane zones of the leeward regions are discussed (Cuddihy 1989).

The original vegetation cover of the Hawaiian Islands prior to the arrival of the Polynesians may be inferred from existing vegetation, remnants in disturbed areas, patterns of climate and substrates, and a few fossilized remains. Forests were and are the natural vegetation of most of the main Hawaiian Islands. On the six Hawaiian Islands with elevations exceeding 1,000 m (3,280 ft), the climate "promotes forest development" except in the alpine zone and the driest parts of the leeward lowlands (Mueller-Dombois 1987). Today, wet forests and shrublands still occur on the windward slopes of the two highest main Islands (Hawai'i, Maui) over a large elevational range and at upper elevations of the islands of Kaua'i, O'ahu, Moloka'i, and Lāna'i. Dry grasslands and shrublands are prominent on the leeward sides of all the Hawaiian Islands; without disturbance, many of these would naturally have been forests. Mesic communities occur on leeward slopes transitional between wet and dry areas, in rain shadows caused by orographic interception, or at elevations above a temperature inversion layer on the higher Islands, usually between 1,525 and 2,135 m (5,000-7,000 ft) elevation (Blumenstock and Price 1967). Subalpine and alpine communities are found on the highest islands of Maui and Hawai'i and are basically dry or mesic types of vegetation. Two of the eight main Islands, Ni'ihau and Kaho'olawe, are here considered "low" islands because they contain only coastal and lowland vegetation. (Lāna'i, lowest of the six main "high" Islands, has a small amount of montane vegetation.) The Northwestern Hawaiian Islands are also low islands; their vegetation is not treated in this volume.

Coastal Vegetation

The coastal zone is a relatively narrow belt encircling each main Island, where vegetation is strongly influenced by the ocean. Coastal communities have been severely altered by humans, and the remaining natural vegetation in this zone is limited in area. Where native plants have not been replaced by alien species, *naupaka-kahakai (Scaevola sericea)* shrubs are often the dominant cover of the strand. Beach morning glory *(Ipomoea pes-caprae)*, beach dropseed *(Sporobolus virginicus)*, *pāʻū o Hiʻiaka (Jacquemontia ovata)*, and *ʻākulikuli* or sea purslane *(Sesuvium portulacastrum)* are also relatively common in remaining strand communities. In some sites, native plants such as *ʻilima (Sida fallax)*, *naio (Myoporum sandwicense)*, *hinahina (Heliotropium anomalum)*, and *nehe (Lipochaeta* spp.) may be locally abundant and even co-dominant with the naupaka. While many remnant strand communities are species poor, the less disturbed coastal sites such as Moʻomomi Beach on Molokaʻi support a rich assemblage of native plants, including very rare species such as *ʻohai (Sesbania tomentosa)*. Vegetation of basaltic shores, cliffs, talus slopes, and coral substrates is also frequently dominated by naupaka (Richmond and Mueller-Dombois 1972), often with the addition of *ʻakoko (Chamaesyce celastroides)*, *maiapilo (Capparis sandwichiana)*, and native sedges and grasses. Many rocky shores have a distinct spray zone community of native sedges and shrubs. On windward Maui and Molokaʻi (at least), this spray zone vegetation consists of low-growing *ʻākia (Wikstroemia* spp.), *ʻakoko*, and *hinahina (Heliotropium curassavicum)* with the sedge *Fimbristylis cymosa*. A species-poor version of this, often supporting only *Fimbristylis*, is common on Hawaiʻi Island.

Coastal shrub communities dominated by natives other than naupaka may be seen as remnants, particularly in less developed, more remote shores of the Islands. Notable among these are shrublands of Hawaiian cotton or *maʻo (Gossypium tomentosum)*, *ʻilima*, and naio, all of which extend somewhat upslope away from the coast. Extrapolating from existing native plants, Char and Balakrishnan (1979) proposed that thickets of naio and the endangered *Achyranthes rotundata* were important in the original vegetation of rocky coral substrates of Barbers Point. It is highly likely that other coastal shrublands existed in pre-human Hawaiʻi, for which remnants no longer exist.

Extant coastal forests occur on some windward shores of Kauaʻi, Maui, Oʻahu, Molokaʻi, and Hawaiʻi. Most often seen are forests dominated by the indigenous *hala (Pandanus tectorius)*. However, the understory of hala forests is usually dominated by Polynesian introductions and alien plants. Atkinson (1970), in a study of succession on dated substrates in the Puna District, island of Hawaiʻi, observed that hala forests were the dominant cover only on flows greater than 200 years in age, although young hala trees could be seen in *ʻōhiʻa (Metrosideros polymorpha)*-dominated vegetation of younger flows. Forests and shrublands of the indigenous *hau (Hibiscus tiliaceus)*, generally mixed with some introduced trees, are seen in wet, sheltered, windward sites (Richmond and Mueller-Dombois 1972). In a few sites on the islands of Molokaʻi and Hawaiʻi, forests dominated by *ʻōhiʻa* and *lama (Diospyros sandwicensis)* occur down to the shoreline. Other coastal areas were forested with *loulu* palms *(Pritchardia* spp.). Remnants of such palm forests may still be seen on Huelo Seastack, an islet off the northern coast of Molokaʻi, as well as on nearby sea cliffs.

Impressions of loulu trunks may be found in the coastal lavas of the Puna District on Hawai'i Island in areas today devoid of such palms. Evidence for the former widespread occurrence of loulu forests has also been found on O'ahu, where fossil remains of the palm were discovered near Salt Lake (Āliapa'akai) (Lyon 1930). Disjunct stands of loulu still grow in a few upslope localities on both O'ahu and Kaua'i. On uninhabited Nihoa, northwest of Kaua'i, a species of loulu (*P. remota*) is a prominent component of vegetation in two valleys of the small island (Conant 1985).

Windward Zones

Prevailing northeasterly trade winds bring moisture-laden air to the northeast-facing slopes of the main Islands. With increasing elevation, air temperature decreases and water precipitates out on windward slopes and some summit areas, resulting in several zones as follows.

Lowland Wet Forests. Wet forests were undoubtedly the predominant original vegetation of the windward lowlands on the larger main Islands (Zimmerman 1948). However, by the late 18th century when European explorers and botanists began to arrive in Hawai'i, the lowland wet zone was primarily a cultivated region (Menzies 1920; Cook 1967). Where lands cultivated by Hawaiians were not subsequently used for agriculture, grazing, or urban development, they were invaded by species of Polynesian introduction, particularly *kukui (Aleurites moluccana)*, or by later introductions such as common guava (*Psidium guajava*). Therefore, most windward valleys and gently sloping tablelands do not support natural forest vegetation, even though many such areas have not been cleared or developed in the last 150 years. The wet lowland zone characterized by Ripperton and Hosaka (1942) makes up over 10% of the five largest Hawaiian Islands, and wet vegetation of all elevations constitutes about one-third of their area. However, because of long-term anthropogenic disturbance, lowland wet vegetation is particularly difficult to reconstruct and characterize, especially on the older, more dissected Hawaiian Islands like O'ahu (Jacobi, in press b).

Natural lowland rain forests may still be seen in regions with rocky substrates or steep terrain. Examples of these forests occur on peaks and lower summit ridges of O'ahu; ridges and *pali* (cliffs) on Kaua'i, Moloka'i, West Maui, and the Kohala Mountains of Hawai'i Island; and in undeveloped slopes of Puna and Hilo Districts on the island of Hawai'i. The most widespread existing lowland forest type is dominated by *'ōhi'a (Metrosideros polymorpha)*, often with an understory of native trees such as *kōpiko (Psychotria* spp.) and *hame (Antidesma platyphyllum)*. The endemic liana *'ie'ie (Freycinetia arborea)* is often abundant in these forests. In the upper reaches of the lowland 'ōhi'a forest on the island of Hawai'i, the understory is dominated by tree ferns or *hāpu'u (Cibotium chamissoi, C. glaucum)*, which form a distinct, closed layer beneath the trees.

A more open 'ōhi'a forest with other scattered native trees and a dense ground cover of the indigenous mat-forming *uluhe (Dicranopteris linearis)* and other related ferns (*Diplopterigium pinnatum, Sticherus owhyensis*) is seen on steep ridges and valley walls of Kaua'i, O'ahu, Moloka'i, Maui, and the Kohala Mountains of Hawai'i Island. This community also covers large expanses of relatively young substrates in the Puna District. Less abundant in Puna is an open 'ōhi'a forest in which *lama (Diospyros sandwicensis)* is a co-dominant. Remnants of this type of forest also occur on steep terrain on windward East Moloka'i, northwestern Kaua'i, and the lower windward slopes of the Kohala Mountains.

8

In pre-Polynesian times, a lowland wet forest dominated by *koa (Acacia koa)* was probably widespread below 1,000 m (3,280 ft) elevation on the larger Islands in windward areas with deep soils. Such a forest still exists in Kīpahulu Valley of windward East Maui, although it probably extended farther downslope before it was subjected to disturbance. Evidence for a lowland koa forest may also be seen on ridges of Hālawa Valley, Molokaʻi (Kirch and Kelly 1975). Skolmen (1986a) speculated that a band of lowland koa forest occurred on the island of Hawaiʻi all along the Hāmākua coast between 305 and 610 m (1,000-2,000 ft) elevation (above the area used for agriculture by the Hawaiians); remnants of this are visible in a few localities.

Montane Wet Forests and Bogs. The boundary between the lowland and montane wet forests in Hawaiʻi is not generally agreed upon by all botanists and ecologists, and it may be variable on the different Islands. In any case, a clear picture of prehuman vegetation is complicated by the extreme disturbance the lowlands have suffered. Jacobi (in press b) used 1,000 m (3,280 ft) elevation to divide the two windward zones on Hawaiʻi Island, but many other classifiers of Hawaiian vegetation place this boundary much lower, generally between 460 m (1,500 ft) and 730 m (2,400 ft) elevation (Hillebrand 1888; Rock 1913; Robyns and Lamb 1939; Ripperton and Hosaka 1942; Krajina 1963).

Wet montane forests still cover relatively large expanses on the islands of Maui and Hawaiʻi and are also found on the steep windward slopes, ridges, and peaks of Kauaʻi, Oʻahu, and Molokaʻi. In his reconstruction of original vegetation, Jacobi (in press b) mapped more than 150,000 ha (370,500 a) of native wet forests above 500 m (1,640 ft) elevation on Hawaiʻi Island alone. Extrapolation from Ripperton and Hosaka's (1942) generalized vegetation maps results in about the same amount of montane rain forest (146,000 ha or 360,620 a) for the islands of Maui, Molokaʻi, Oʻahu, and Kauaʻi combined. Most of these forests are dominated by *ʻōhiʻa (Metrosideros polymorpha)* in a closed canopy with a well-developed understory of mixed native tree species, shrubs, and tree ferns (*Cibotium* spp.). Tree ferns often comprise a very dense layer in some montane forests of Hawaiʻi Island.

ʻŌlapa (Cheirodendron spp.) is often co-dominant with ʻōhiʻa in low-statured, wind-stunted forests of peaks and ridges, areas almost continually brushed by clouds. These cloud forests occur on Kauaʻi, Oʻahu, Molokaʻi, West Maui, and in the Kohala Mountains of Hawaiʻi. A dense cover of gnarled trees, abundant epiphytes (particularly mosses and liverworts), and a diversity of small native trees, shrubs, and ferns are characteristic of this vegetation (Fosberg 1972). On the older islands such forests may contain several species of *ʻōhiʻa (Metrosideros polymorpha, M. tremuloides, M. rugosa)*.

On other steep montane slopes, the ʻōhiʻa forest is more open, and *uluhe (Dicranopteris linearis)* and related ferns are the ground cover. On the island of Hawaiʻi, wet montane forests (above 1,000 m) with a significant uluhe component cover only 4,000 ha (9,880 a); this type of vegetation is much more common at lower elevations (Jacobi, in press b).

Less common are wet montane forests with tall *koa (Acacia koa)* emergent above a closed canopy of ʻōhiʻa. For the island of Hawaiʻi above 500 m (1,640 ft) elevation, Jacobi (in press b) mapped wet koa/ʻōhiʻa forests over less than 35,000 ha (86,450 a). This type of forest may once have occurred in a discontinuous band on windward slopes of Mauna Kea, Hualālai, and both windward and leeward slopes of Mauna Loa, as well as on the windward slopes of Haleakalā, East Maui (Skolmen 1986a). While much reduced through clearing, logging, and ranching activities, a few good examples of this species-rich type of rain forest remain on both Maui and Hawai'i (Smathers 1967; Cooray 1974).

Wet shrublands of the montane zone are found on steep valley walls, cliffs, and ridge crests of all the higher Islands. These may be dominated by stunted 'ōhi'a and other components of nearby forests, or on very steep slopes by native ferns, particularly 'ama'u (*Sadleria* spp.).

Another type of montane wet community is represented by bogs, which occur on most of the high Hawaiian Islands. Bogs, generally small in area, are found in very wet, poorly drained places near mountain summits on Kaua'i, O'ahu, Moloka'i, and West Maui (Selling 1948; Carlquist 1980); on high-rainfall windward slopes of East Maui (Loope et al., in press a); and on windward slopes of the Kohala Mountains, Mauna Kea, and Mauna Loa on Hawai'i Island (Jacobi, in press b). They are characterized by sedges and grasses (*Oreobolus furcatus, Carex* spp., *Rhynchospora* spp., *Dicanthelium* spp.) and stunted woody plants including 'ōhi'a, *pilo (Coprosma ochracea)*, and *na'ena'e (Dubautia* spp.). A diversity of unusual bog-tolerant plants occurs here: greenswords *(Argyroxiphium grayanum)*, lau-kāhi *(Plantago* spp.), the diminutive, daisy-like *Lagenophora viridis*, the striking, fleshy-trunked lobelias of Maui and Kaua'i (*Lobelia gloria-montis, L. kauaensis, L. villosa*), the fragile clubmoss *Selaginella deflexa*, and endemic violets *(Viola maviensis, V. robusta*, and others). Each bog, however, is distinct and contains a unique assemblage of native plants. Bogs of the Alaka'i Swamp on Kaua'i were recently studied by Canfield (1986), who recognized three distinct plant communities based on species composition, life form, and structure. A comprehensive and detailed description of the botanically rich bogs of East Maui has been completed by Loope et al. (in press a). One unique high-elevation East Maui bog surrounded by *Deschampsia* grassland was characterized as "alpine" by Vogl and Henrickson (1971) but has much in common with its lower-elevation neighbors.

In addition to small open or raised bogs, the windward slopes of Mauna Kea on the island of Hawai'i contain large expanses of a very wet 'ōhi'a community that has been described as "bog-formation dieback." In very old, water-soaked soils, stands of 'ōhi'a trees undergo dieback and are replaced by low 'ōhi'a shrubs and sedges (Mueller-Dombois 1986).

Based on pollen analysis from upland bogs, Selling (1948) determined that the area covered by rain forest in the montane zone fluctuated with climatic change in the late quarternary period. Selling recognized three periods prior to the present, with the greatest extension of the montane rain forest occurring during the second period. The most important rain forest constituents of this period (based on pollen samples) correspond well with typical dominants found at present in montane wet forests: *Metrosideros, Myrsine, Cheirodendron, Cibotium*, and *Coprosma*. Selling's third period, in which the rain forest retreated and was replaced in some areas by drier vegetation types and plants more typical of subalpine areas, corresponds to the period of Polynesian occupation of the Islands.

Montane Moist Forests and Parkland. Moist or mesic forests occur in the upper montane zone on the islands of Maui and Hawai'i, where they are found on the edges of orographic rain shadows, on windward slopes of mountains near the temperature inversion layer, and on upper leeward slopes. These forests differ from wet forests in the relative scarcity of tree ferns (*Cibotium* spp.) and epiphytes, the abundance of shrubs such as *pūkiawe (Styphelia tameiameiae)* in the understory, and a different complement of native ferns in the ground cover. For most of these forests the dominant trees are either *'ōhi'a (Metrosideros polymorpha), koa (Acacia koa),* or a mixture of these two species. In a very few sites, *mānele (Sapindus saponaria)* is a co-dominant species in the koa/'ōhi'a canopy. Montane mesic forests have a very restricted distribution and together amount to just slightly more than 50,000 ha (123,500 a) above 1,000 m (3,280 ft) on the island of Hawai'i; the koa/'ōhi'a mixed forests are important endangered bird habitat and are extremely vulnerable to disturbance (Jacobi, in press b). Maui and Kaua'i each have only a few thousand hectares of mesic montane forest (Ripperton and Hosaka 1942).

Drier parklands or open woodlands of low, spreading koa and *māmane (Sophora chrysophylla)* trees with native shrubs and grasses were formerly common in the transition area between wet forests and the subalpine zone, at least on the island of Hawai'i. Protected examples of such "mountain parklands" have been described in detail by Robyns and Lamb (1939) and Mueller-Dombois (1966, 1967). Jacobi (in press b) has mapped approximately 30,000 ha (74,100 a) of this type of vegetation on the island of Hawai'i.

Subalpine Vegetation

The high mountains of East Maui (Haleakalā) and Hawai'i (Mauna Kea, Mauna Loa, Hualālai) support extensive native subalpine ecosystems above 2,000 m (6,560 ft) elevation. Most of the plant communities of this zone are dry; grasslands, shrublands, and forests are all represented here. Subalpine grasslands are dominated by the endemic bunchgrass *Deschampsia nubigena*. In cindery substrates such as the upper slopes of Mauna Kea and within Haleakalā Crater, grasslands are relatively dry. More mesic examples of *Deschampsia* grassland occur on the windward slopes of Haleakalā, which are frequently covered by moisture-bearing clouds.

Dry shrublands of the upper mountain slopes are most often dominated by *pūkiawe (Styphelia tameiameiae)* and *'ōhelo (Vaccinium reticulatum)*, but *'a'ali'i (Dodonaea viscosa)* and *na'ena'e (Dubautia ciliolata)* may also be important components. Frequently, very scattered *'ōhi'a (Metrosideros polymorpha)* trees are found here. A distinctly different dry shrubland composed primarily of *'āweoweo (Chenopodium oahuense)* occurs on ashy substrates in the saddle between Mauna Loa and Mauna Kea on Hawai'i Island. This shrubland may, at least in some areas, represent former subalpine forest degraded by feral animals, fire, and the activities of humans at a nearby military training camp.

Mesic shrublands dominated by *'ōhelo (Vaccinium* spp.) and *'ama'u* fern (*Sadleria cyatheoides*) are found on the steep upper slopes of East Maui, sandwiched between wet forests and mesic grasslands. The Haleakalā Crater area supports a community of very scattered silversword plants (*Argyroxiphium sandwicense* subsp. *macrocephalum*) and *na'ena'e (Dubautia menziesii)*, primarily on cinder cones (Loope and Crivellone 1986). A similar community once occurred on Mauna Kea, where silverswords (subsp. *sandwicense*) were so common that the dry leaves and stems were used as fuel for campfires (Douglas 1914). Feral ungulates have nearly exterminated the silversword on Mauna Kea.

Forests of the subalpine zone are relatively dry; two main types of subalpine forest exist. Open forests of low-statured 'ōhi'a are widespread on less weathered, younger substrates of Hawai'i Island, while open to closed forests of *māmane (Sophora chrysophylla)* occur on cinder, ash, and old weathered lavas of both Maui and Hawai'i. In some māmane forests of Hawai'i Island, especially on Mauna Kea and in the saddle area, *naio (Myoporum sandwicense)* is a co-dominant in a low-statured woodland, called by Robyns and Lamb (1939) "plateau parkland." In some areas, naio grows with māmane all the way to the treeline; the community also extends below the subalpine zone. Subalpine māmane forests have been mapped over nearly 18,000 ha (44,460 a) of Hawai'i Island, those with naio cover another 6,000 ha (14,820 a) in the subalpine zone, and an equal area in the montane zone (Jacobi, in press b).

Alpine Vegetation

Vegetation above 2,800 m (9,180 ft) is very sparse and is composed almost entirely of native plant species. Commonly seen shrublands of alpine Maui and Hawai'i are composed of very scattered low *pūkiawe (Styphelia tameiameiae)* and *'ōhelo (Vaccinium reticulatum)*. Very high-elevation areas of Hawai'i Island near the summits are essentially devoid of plants and support only a few native lichens, mosses (particularly *Racomitrium lanuginosum*), and grasses (*Trisetum glomeratum, Agrostis sandwicensis*) (Mueller-Dombois and Krajina 1968). Subalpine and alpine communities were probably affected very little by Hawaiians, and even today they remain largely undeveloped, except for buildings such as astronomy observatories and their attendant roads. However, even these high-elevation areas have been impacted by feral and domestic ungulates.

Dry Leeward Zones

The leeward regions of the Hawaiian Islands are much drier than windward-facing slopes, which intercept the moisture-bearing trade winds. Except for parts of the Kona slopes of Hawai'i Island, the leeward sides of the Islands typically receive less than 1,250 mm (50 in.) of annual precipitation, and some areas have less than 250 mm (10 in.) per year. Dry and mesic leeward communities begin directly below the subalpine zone on the high islands of Maui and Hawai'i, and below the summit rain forests on Kaua'i, O'ahu, and Moloka'i. The lower islands of Ni'ihau, Kaho'olawe, and Lāna'i are located in rain shadows of larger islands, and their vegetation may be considered to belong almost entirely to the dry leeward zone. Perhaps because of a history of human disturbance, the vegetation of the dry leeward zone is more fragmented and difficult to characterize than that of wet windward zones.

Lowland Grasslands and Shrublands. Grasslands, dominated by alien grass species, are widespread today in the lowlands of all the Hawaiian Islands and in some areas extend above 610 m (2,000 ft) elevation. A few lowland areas on the islands of Hawai'i, Maui, Lāna'i, Kaho'olawe, and Moloka'i support grasslands of *pili (Heteropogon contortus)*, a species either indigenous to the Islands or introduced by Polynesians. Extensive lowland grasslands, most likely of pili, were noted by some of the early European travellers to Hawai'i (Ellis 1827; Beaglehole 1967; Macrae 1972) but were probably largely the result of the Hawaiian practice of burning. Such grasslands may have been entirely anthropogenic in origin (Kirch 1982). It is likely that endemic annual grasses of the genus *Panicum*, such as *kākonakona (P. torridum)*, were formerly more important components of dry lowland vegetation, where they probably grew intermixed with native shrubs. Grasslands dominated by the indigenous *'emo-loa*

(*Eragrostis variabilis*) were probably more extensive in pre-human Hawai'i; remnants may be seen today on cliffs and steep valley walls on several Islands, most notably Kaua'i, O'ahu, and Moloka'i. On southeastern O'ahu, *Eragrostis* grasslands are found on steep east- or northeast-facing slopes exposed to strong winds, where the grass grows with other natives such as *nehe (Lipochaeta integrifolia)* and the sedge *Fimbristylis cymosa* (Kartawinata and Mueller-Dombois 1972). *Eragrostis* is also an important component of the vegetation of some of the Northwestern Hawaiian Islands (Christophersen and Caum 1931; Conant 1985; Herbst and Wagner, in press).

Native shrublands and forests were undoubtedly much more characteristic of the low-land vegetation before the advent of humans in Hawai'i and probably extended to the coast in many places (Ripperton and Hosaka 1942; Zimmerman 1948). On some leeward slopes of the Islands and on more recent substrates of windward Hawai'i Island, dry and mesic shrublands may still be seen in areas where agriculture and grazing have not been intensive. Native shrubs that are dominants in these communities are *'a'ali'i (Dodonaea viscosa)*, *'ākia (Wikstroemia spp.)*, *'āweoweo (Chenopodium oahuense)*; *ko'oko'olau (Bidens menziesii)*, *pūkiawe (Styphelia tameiameiae)*, *alahe'e (Canthium odoratum)*, and low-growing *'ōhi'a (Metrosideros polymorpha)*. Such shrubs usually occur in mixed stands with two or more co-dominant shrubs; nonnative grasses are now important components of these shrublands. On steep cliffs, other shrubs such as *'akoko (Chamaesyce spp.)*, *nehe (Lipochaeta spp.)*, and *kulu'i (Nototrichium sandwicense)* may represent remnants of formerly more extensive communities. Two very rare shrublands extant in only limited areas give some indication that other shrub communities have disappeared with disturbance; these are a very open community of *'ohai (Sesbania tomentosa* f. *arborea)* and other natives on the leeward slopes of Moloka'i, and a shrubland in the Kōke'e and Nā Pali regions of Kaua'i containing *iliau (Wilkesia gymnoxiphium)* among more common shrubs such as pūkiawe.

Lowland Dry and Mesic Forests. Lowland leeward forests were considered by Rock (1913) to be the richest of all Hawaiian forests in terms of numbers of tree species and unique plants, but today they have been reduced to mere remnants over much of their original range. Most remaining native dry and mesic lowland forests are dominated by *lama (Diospyros sandwicensis)* or *'ōhi'a (Metrosideros polymorpha)*. Either *'ōhi'a* or lama may predominate, but often they occur together in mixed stands. Remaining lama/'ōhi'a forests are, for the most part, found on very rocky substrates, steep slopes, or in gulches, areas unsuitable for agricultural development or clearing. In drier, leeward areas of the island of Hawai'i, these forests may contain *alahe'e (Canthium odoratum)*, *kauila (Colubrina oppositifolia)*, *'aiea (Nothocestrum breviflorum)*, *wiliwili (Erythrina sandwicensis)*, *'ohe makai (Reynoldsia sandwicensis)*, and endangered tree species such as *uhiuhi (Caesalpinia kavaiense)* and *koki'o (Kokia drynarioides)*. Mesic lama/'ōhi'a forests may still be seen in the Puna and South Kona Districts of the island of Hawai'i. Of all the lowland dry forests surveyed by Jacobi (in press b) on the island of Hawai'i (500-1,000 m or 1,640-3,280 ft), only those dominated by 'ōhi'a had significant cover (> 20,000 ha or 49,400 a).

More rarely-seen lowland forests are those in which the predominant tree is wiliwili, *koa (Acacia koa)*, *koai'a (A. koaia)*, *olopua (Nestegis sandwicensis)*, or *āulu (Sapindus oahuensis)*. The āulu forest type is seen only on O'ahu, where it is restricted to lower slopes and gulch walls of the leeward Wai'anae Mountains. Hatheway (1952) suggested that *Sapindus* and lama forests once covered much of the lower forested region of leeward O'ahu. Wirawan (1974), who two decades later studied the same forest stands as Hatheway, noted that some native dry forest species were being suppressed by alien grasses and shrubs, but that *Sapindus* was able to maintain dominance and even invade nonnative stands. *Sapindus*-dominated forests may formerly have occurred on Kaua'i, in Waimea Canyon and Makaweli, where Rock (1913) noted that the species was scattered among *kukui (Aleurites moluccana), hōlei (Ochrosia* sp.), and *kōpiko (Psychotria* spp.).

Open forests or savannas of wiliwili must have been very widespread on lower leeward slopes, where scattered trees and stands are still commonly seen on undeveloped lands or rough substrates. One of the best remaining examples of a wiliwili forest was described by Medeiros et al. (1984) from a site between 200 and 400 m (660-1,310 ft) elevation above Kīhei, Maui. More than a dozen native trees and shrubs were components of this forest, including such rarities as koai'a, *ko'oloa'ula (Abutilon menziesii)*, and the native yellow *Hibiscus* or *ma'o hau hele (H. brackenridgei)*. This exceptional dry forest remnant also contained a rich herbaceous flora including native grasses (*Panicum* spp.), ferns, herbs, and vines (*Sicyos* sp., *Ipomoea* spp., *Bonamia menziesii, Canavalia haleakalensis*).

Based on remnants of native vegetation, Char and Balakrishnan (1979) speculated that the original vegetation of the 'Ewa Plains of O'ahu was an open wiliwili savanna with other trees such as āulu, sandalwood (*Santalum ellipticum* and *S. freycinetianum*), and naio (*Myoporum sandwicense*). Native shrubs, including 'akoko (*Chamaesyce skottsbergii*), ma'o (*Gossypium tomentosum*), *Achyranthes rotundata, Abutilon incanum*, and 'ilima (*Sida* spp.), as well as vines, grasses, and a few ferns were also part of this community. Supporting evidence for an open native forest such as this comes from a study of the subfossil land snails of the area (Christensen and Kirch 1981). On Kaho'olawe, extant wiliwili trees on the hill of Moa'ula, evidence from archaeological studies (TenBruggencate 1986a), and 19th century reports of plant cover (Myhre 1970) are indications that the Island probably supported a forest similar to that described for the 'Ewa Plains.

Koai'a forests have been almost entirely converted into pastures. Only in areas fenced to exclude cattle or in inaccessible gulches may remnants of this rare community be seen. In a botanical survey of the Kawaihae-Waimea area on Hawai'i Island, McEldowney (1983) found relict koai'a forests in pastureland ravines between 610 and 1,190 m (2,000-3,900 ft) elevation. Trees associated with koai'a near Kawaihae are naio, olopua, *māmane (Sophora chrysophylla)*, sandalwood (*Santalum* sp.), and wiliwili. Although not strictly a forest type on Maui, koai'a was found to be associated with many native woody species in several stands on the south slope of Haleakalā (Medeiros et al. 1986).

A few forests dominated by olopua or by olopua and lama may still be seen on Lāna'i, Maui, and Hawai'i. The Lāna'i example was described in detail by Spence and Montgomery (1976) and is notable for the number of rare plants it contains. Two federally listed endangered tree species grow there: the Hawaiian gardenia or *nā'ū (G. brighamii)* and the Lāna'i sandalwood (*Santalum freycinetianum* var. *lanaiense*) (U.S. Fish and Wildlife Service 1987). Spence and Montgomery speculated that this type of forest may have once covered much more area on Lāna'i, before the introduction of grazing and browsing animals and subsequent severe wind erosion. Ziegler (1989), who also studied the Kānepu'u forest, estimated that half of Lāna'i could have originally supported a dry forest or woodland.

In a very few sites on leeward slopes of Kaua'i, O'ahu, West Maui, and Lāna'i are remnants of extremely species-rich dry to mesic forests with no clear dominant tree species; these are refugia for nearly extinct trees such as *mēhamehame (Flueggea phyllanthoides)* as well as other rare plants such as *a'e (Zanthoxylum* spp.), *māhoe (Alectryon macrococcum)*, and *halapepe (Pleomele* spp.). Several botanically diverse dry forests of Maui (south slope of Haleakalā) and Hawai'i (Pu'uwa'awa'a and Kapu'a) are described in detail by Rock (1913) and Judd (1932). It was Rock's opinion that a band of this rich dry forest once "encircled the southern slopes of Mauna Loa" but was destroyed by lava flows. Even the relatively young substrates of the East Rift of Kīlauea support dry and mesic forests in areas not overrun by lava for several hundred years (Mueller-Dombois 1966; S.J. Anderson et al., unpubl. data).

Lowland dry and mesic forests dominated by koa, āulu, and olopua, as well as the extremely rich mixed forests, occur as small remnant stands; over most of their original range they have been replaced by alien trees and shrubs. In some leeward lowland areas such as northwestern Kaua'i and Ni'ihau, the only indications of the former native forest cover are the subfossil remains of endemic land snails of the genus *Carelia* (Gage 1988).

Some areas that are quite xeric today may have been more mesic in the past. Fossils of native trees buried in ash near Salt Lake on O'ahu dated from 100,000 years ago indicate that the forest in this leeward area near sea level contained native tree species such as 'ōhi'a, koa, *loulu* palm *(Pritchardia* sp.) (Lyon 1930), and *Pteralyxia* sp. (Selling 1948), species found on O'ahu today only in moister sites at higher elevations.

Montane Grasslands and Shrublands. Grasslands are not an important element in native Hawaiian montane vegetation between approximately 500 and 2,000 m (1,640-6,560 ft). The extensive pasturelands now found in the montane zone of Maui and Hawai'i are not natural but are converted forests and shrublands. In a few dry and mesic areas of Hawai'i Island on the slopes of Mauna Kea, Mauna Loa, and the saddle between them are grasslands dominated by endemics such as *Eragrostis atropioides, Deschampsia nubigena*, and *Panicum tenuifolium*. These grasslands may have covered more area before the advent of alien grasses and introduced ungulates, or they may represent degraded forests.

Shrublands are also not prominent in the leeward montane zone. Dry shrublands containing *'a'ali'i (Dodonaea viscosa), na'ena'e (Dubautia linearis)*, and *ko'oko'olau (Bidens menziesii)* occur on leeward slopes of Mauna Loa and the saddle region among the mountains; elsewhere on Mauna Loa *pūkiawe (Styphelia tameiameiae)* is an important shrubland plant. A dry shrubland composed mainly of low, scattered *'ōhi'a (Metrosideros polymorpha)* and native shrubs is found on leeward slopes of Hawai'i and Maui and may represent an early successional stage of dry 'ōhi'a forest. Jacobi (in press b) mapped this dry shrub community over nearly 20,000 ha (49,400 a) of Hawai'i Island's montane zone.

Montane Dry Forests. Forests are the primary natural vegetation of the leeward montane zone. Dry forests are found on leeward East Maui and Hawai'i and are dominated by one or more of the following tree species: *koa (Acacia koa), māmane (Sophora chrysophylla)*, the ubiquitous *'ōhi'a (Metrosideros polymorpha)*, and much more rarely, *'akoko (Chamaesyce olowaluana, C. celastroides)*. Particularly rich examples of montane dry forests occur at Auwahi on Maui and Pu'uwa'awa'a on the slopes of

Hualālai, Hawai'i; these are described in detail by Rock (1913). More mesic sections of these rare forests have *olopua (Nestegis sandwicensis)* as an important component. While changes (particularly the invasion of alien grasses) have occurred in the 75 years since Rock's work in these areas and may have resulted in drier conditions, both sites remain notable for great tree species diversity and the presence of rare plants (Powell and Warshauer 1985a; Medeiros et al. 1986). Based on the current and historical distribution of tree species characteristic of the Auwahi forest, Medeiros et al. (1986) speculated that a rich dry forest once had a much greater distribution on the south slope of Haleakalā.

* * *

Major activities of humans in modifying Hawai'i's native vegetation will be discussed in the rest of this volume according to time periods, islands, and vegetation zones. A summary of the changes and an indication of what natural areas remain conclude the review.

EFFECTS OF ABORIGINAL PEOPLE
ON VEGETATION OF THE HAWAIIAN ISLANDS

The Hawaiian Islands were settled by ocean-voyaging Polynesians, probably from the Marquesas, an island group 4,000 km (2,400 mi) to the southeast. Evidence for the affinity of the Hawaiian and Marquesan peoples is derived from similarities of language, body type, and agricultural plants (Kirch 1985a). However, there may have been more than one wave of immigration, and contact between Hawai'i and island groups other than the Marquesas (i.e., the Society Islands) apparently occurred as well (Dye, in press).

The time of first settlement has been variously reported. Earlier estimates were often around 800 A.D. (Green 1971; Handy and Handy 1972), but Kirch (1985a) estimated that colonization of the Hawaiian Islands occurred as early as 300 A.D., or more conservatively 400 A.D. (Kirch 1982). Kirch reasoned that such an early date for first arrival is likely because permanent settlements were distributed on all the main islands by the 6th century and such development would have required time. One of the earliest dated archaeological sites that has been excavated is the Bellows or Waimanalo Beach site on O'ahu, which appears to have been initially occupied between 450 and 500 A.D. (Kirch 1974) or possibly as early as 327 A.D. (Tuggle et al. 1978). Another site of early human habitation is Hālawa Dunes on Moloka'i, dated from 600 A.D. (Kirch 1974; Kirch and Kelly 1975).

Human Population Size

Although humans have been occupying the Hawaiian Islands for more than 1,500 years, the size of the original population was probably quite low, and the impact on the Hawaiian environment was correspondingly small at first. The original group settling the Islands may have been as few as 100 people (Kirch 1982). This population built up relatively slowly, and by the end of the colonization period (600 A.D.) may have amounted to only 1,000 people (Kirch 1985a). Then ensued a period of development, lasting until about 1100 A.D., in which the population increased and expanded outward until settlements were scattered throughout the lowlands of the main Hawaiian Islands. There were still lands available for agriculture and ample resources to be exploited (Kirch 1985a). Even in favorable localities, such as Hālawa Valley, Moloka'i, human populations may have remained relatively low until the 13th or 14th century (Kirch and Kelly 1975).

After about 1100 A.D., the Hawaiian population began to increase dramatically, on the western part of Hawai'i Island doubling every century (Kirch 1985a), and agricultural systems were expanded and intensified (Rosendahl 1974). A population peak (usually estimated at several hundred thousand) was reached around 1650 A.D., more than 100 years before contact with Europeans (Kirch 1982; Kelly 1983). It was at this population peak, or shortly before, that Hawaiians began to inhabit less favorable coastline areas (Griffin et al. 1971) and barren zones between the coast and upland agricultural sites (Rosendahl 1973) and to develop extensive dryland agricultural systems in marginal regions (Kirch 1985a).

Large-scale irrigation works and permanent field systems were developed during the expansion period. Population densities in the fertile windward valleys may have approached 250 people/km^2 (Kirch 1982), although densities in tablelands and elsewhere would have been much lower (Newman 1969). There is evidence from archaeological studies on West Hawaiʻi that after 1650 A.D. the Hawaiian population declined (Hommon 1976), or at least the rate of growth greatly decreased. In some marginal habitats such as Kahoʻolawe, the human population may have declined even earlier (by 1550 A.D.), probably because of environmental degradation caused by clearing natural vegetation (Hommon 1980). Assuming that the population of the Hawaiian Islands was declining more than 100 years before contact with Europeans, Kirch (1982) speculated that the ability of the Islands to support a high human population had been surpassed and that environmental deterioration was occurring, reducing the land's carrying capacity. Abandonment of some dry coastal sites, such as Waiʻahukini in the Kaʻū District (Emory et al. 1969) and Kalāhuipuaʻa in the South Kohala District of Hawaiʻi Island (Kirch 1979), occurred in this late prehistoric period.

The population of the Hawaiian Islands at the time of initial European contact has been variously estimated at 200,000 (Schmitt 1971) to greater than 1 million (Stannard 1989). Captain James King, who was on Cook's voyage to Hawaiʻi in 1778-79, used settlements along Kealakekua Bay to project a population of 150,000 for the island of Hawaiʻi alone and calculated a total of 400,000 people for all the Hawaiian Islands (Schmitt 1968). Stannard (1989) based his population estimate of 800,000 to 1 million on a re-evaluation of estimates and assumptions of early European visitors to Hawaiʻi and on a comparison with accepted population estimates in similar areas of Eastern Polynesia.

Stannard (1989) speculated that a great decrease in the Hawaiian population began immediately after the first arrival of Europeans (1778), who brought with them diseases such as syphilis, tuberculosis, and influenza. Since the isolated Hawaiian people had no acquired immunity to these and other diseases, the result was a devastating loss of perhaps more than half the pre-European contact population in less than 25 years. Stannard postulated a population decline ratio of 20 to 1 within 100 years of western contact.

Missionaries began taking censuses of the Hawaiian population in the 1830s, and by the 1850s the records indicated a dramatic decline. The district of Kaʻū on Hawaiʻi Island, for example, lost half its population between 1835 and 1853, and the censused number of people in Kaʻū in 1853 is thought to be only about 15% of the projected population in 1778 (Kelly 1969). Schmitt (1968) estimated that the population of the island of Hawaiʻi decreased from 120,000 in 1779 to 46,000 in 1831 because of diseases and disruption of the Hawaiian economic system. With the change to a money economy, centers of population shifted, and many formerly productive agricultural areas and fishing villages were depopulated or completely abandoned. For example, during early historical times, Hālawa Valley on Molokaʻi supported 400 people (Griffin et al. 1971); today it is almost uninhabited. Likewise, fishing villages all along the Kaʻū coast, such as Kealakomo where William Ellis reported more than 300 people gathering to hear a sermon in 1823 (Ellis 1827), were deserted by the end of the 19th century.

Agriculture

Kirch (1985a) called vegetation alteration for agriculture "the greatest force leading to environmental change in pre-contact times." Permanent agriculture in Hawaiʻi took two main forms: wetland taro or *kalo (Colocasia esculenta)* cultivation, and dryland cultivation of sweet potato (*Ipomoea batatas*), taro, and other crops. Irrigation systems in valleys of the Hawaiian Islands were both intensive and sophisticated; they are among the largest recorded in Polynesia (Kirch 1985a). In essentially every valley with a permanent stream, taro fields were irrigated (Handy and Handy 1972; Kirch 1982). Irrigation systems varied in size and complexity in the different valleys of the islands, but most were composed of pondfields or terraces (*loʻi*) constructed of stone, and ditches or *ʻauwai*. The earliest dated irrigation system is that of Hanalei, Kauaʻi, estimated at 600 A.D. (Kirch 1982). There is some disagreement among archaeologists about the meaning of this early date for Hanalei. Some believe that the 7th century date represents habitation in the area before pond and terrace construction, and that large-scale irrigation works were developed much later, in the 13th to 15th centuries (Kirch 1985a). Occupation of other nearby Kauaʻi sites has been dated to the 10th century, with speculation that intensive irrigation agriculture did not begin until the 15th century (Hammatt et al. 1978; Kirch 1985a). Many other valleys on Oʻahu and Molokaʻi were first irrigated between 1400 and 1500 A.D. (Yen et al. 1972; Kirch and Kelly 1975). While wet cultivation of taro probably began with the earliest Hawaiian colonizers, major irrigation works were relatively late developments dating from the 15th to the 19th century (Kirch 1985b).

Valleys of windward Kauaʻi, Molokaʻi, Maui, and Kohala District on Hawaiʻi Island, as well as both windward and leeward valleys of Oʻahu, show abundant evidence of Hawaiian irrigation farming (Kirch 1985a). Some sites, such as Keʻanae on Maui and Waipiʻo on Hawaiʻi, have never ceased taro cultivation. However, some of the most productive taro lands today (in the lowest alluvial plains of valleys) were only marginal or not used by prehistoric Hawaiians because of limitations in their drainage technology (Spriggs 1985). Since it was only in the 1960s that remains of agricultural systems were recognized as important archaeological sites, many important agricultural areas of the past have been urbanized and developed, such as Mākaha, Oʻahu (Yen et al. 1972).

Because of the great production of food available from irrigated taro fields, or perhaps because of the number of people required to construct and care for such irrigation systems, important sites of irrigation farming were also centers of population in ancient Hawaiʻi. On Kauaʻi, these centers were located at Kapaʻa, Nāwiliwili Bay, Waimea, Hanapēpē Valley, and along the Nā Pali coast. Oʻahu, which had the greatest amount of irrigated taro-producing land in the Islands, had its densest human population at Waikīkī and the large valleys nearby. Other heavily populated irrigated taro lands were found along the North Shore and in windward valleys between Lāʻie and Kāneʻohe. The most populated areas of Maui were streams between Kahakuloa and Waikapū, including Wailuku; Lahaina and adjacent small valleys; and the northern shore of East Maui near Keʻanae and Hāna (Handy and Handy 1972). Kahikinui, on the south slope of Haleakalā, apparently supported a large population in both coastal and upland agricultural zones (Kirch 1985a). Molokaʻi inhabitants were concentrated in the windward valleys of the eastern half of the island, particularly Hālawa, Wailau, Pelekunu, and Waikolu (Kirch 1985a). Excepting Waikīkī, Kāneʻohe, Lahaina, and Wailuku, most of these early centers of habitation do not correspond with modern cities.

The smaller islands of Niʻihau and Lānaʻi, as well as western Molokaʻi, had no wetland taro cultivation, and the relatively low human populations of these areas depended on the sweet potato and yam (*Dioscorea alata*) as major crops (Handy and Handy 1972). The sweet potato or *ʻuala* was the most important crop of dryland cultivation systems because of its high yields under dry conditions and the ease with which it could be propagated (Yen 1974; Kirch 1985b). The uplands of Kahoʻolawe were apparently cultivated in the past, but dryland farming ceased there long before Europeans arrived to record it (Hommon 1980; Kirch 1985a).

Hawaiʻi Island may be used as an example of the sequence of Hawaiian agricultural development, as much archaeological work has been done there. The first humans to colonize the island of Hawaiʻi probably settled initially along the Kona coast and in valleys of the eastern part of the Kohala District (Newman 1969). West Hawaiʻi was settled by 750 A.D., if not earlier (Kirch 1974). The general pattern for settlement in North Kona began with the occupation of the coast, where the exploitation of marine resources was more important than farming. Later, the center of habitation moved upland to an area with greater rainfall, and extensive agriculture was the main way of life (Rosendahl 1973). The farmers of the uplands then traded with the fishing communities, exchanging agricultural produce for fish and shellfish.

Kirch (1985a) outlined the sequence of agricultural development and the corresponding population growth and cultural development on the island of Hawaiʻi. When the population was small, agriculture was primarily shifting cultivation (slash and burn or swidden) with long fallow periods. As the human population increased, the amount of area cultivated increased, cultivators moved upland, and the fallow period between crops decreased. Eventually, permanent agriculture displaced swiddening over most of the Island (Newman 1969).

Hawaiʻi Island Field Systems. Two main permanent agricultural zones existed on the island of Hawaiʻi: the eastern valleys of the Kohala District, in which irrigation was practiced on a large scale, and tablelands, which were primarily sites of dryland farming. The dry farming areas may be further divided into scattered fields of the lower windward slopes of the Hāmākua, Hilo, and Puna Districts, and large field systems of Kohala, Waimea, Kona, and Kaʻū (Newman 1969). The four large field systems occurred in the only regions of leeward Hawaiʻi with soils suitable for agriculture and annual rainfall greater than 500 mm (20 in.) (Kirch 1985a). These leeward sites also originally supported open vegetation more amenable to clearing (with stone tools and slash and burn techniques) than the dense rain forests of the windward slopes. Sophisticated dryland field systems were developed primarily on leeward Hawaiʻi, but also on leeward East Maui, and were intensifications or extensions of past shifting agriculture (Kirch 1985a). The areas supporting the large field systems, particularly Kona and Kaʻū, which also had rich fishing grounds offshore, were centers of population on the island of Hawaiʻi.

<u>Kona Field System.</u> The Kona field system was the most highly developed agricultural area of Hawaiʻi Island (Handy and Handy 1972) and may have supported more than half the total population of the island at the time of European contact (Kelly 1983). Although it has not been completely mapped, the Kona field system was a band 5 km (3 mi) wide and 29 km (18 mi) long, located above Kailua and Kealakekua (Newman 1970). This single cultivated area covered an estimated 139 km^2 (56 mi^2) (Kirch 1985a), more than 1% of the entire land area of the island of Hawaiʻi. This is even more significant when one considers that greater than 86% of the Island's area is currently considered unsuitable for agriculture (Schmitt 1989). The best-preserved portions of the system are in the uplands of Kealakekua (Kirch 1985a), where stone and earth walls run both perpendicular and parallel to the slope, defining long,

rectangular fields. The Kona field system is depicted in early drawings of Kailua by the missionary William Ellis and by Lucy Thurston (Ellis 1827; Kelly 1983).

Four distinct zones of cultivation at Kona were described by Kelly (1983). The first was the *kula* lands, the dry coastal plain up to about 150 m (500 ft) elevation. Where soil was sufficient, sweet potatoes (*Ipomoea batatas*) and *wauke* (*Broussonetia papyrifera*) were cultivated. The kula zone was also the area in which fields of *pili* grass (*Heteropogon contortus*), used for thatching by Hawaiians, were encouraged through the use of fire (Kirch 1985a). Above this, up to about 300 m (1,000 ft) elevation, was the *kaluʻulu* zone, where plantations of *ʻulu* or breadfruit (*Artocarpus altilis*) were prominent, but sweet potatoes and wauke were also planted, apparently among the trees (Kelly 1983). The botanist Menzies (1920), passing through this zone above Kealakekua in 1793, described the kaluʻulu zone in some detail, commenting on the beauty of the breadfruit trees and the industry displayed in Hawaiian cultivation. More than any other aspect of the Kona field system, intensive breadfruit plantations are indicative of organization and central planning and suggest participation of some central authority in their development (Kelly 1983).

At an elevation above the kaluʻulu zone, from 305 to 760 m (1,000-2,500 ft), was the *ʻapaʻa* zone, perhaps the most intensively cultivated and productive of all the zones (Kelly 1983). Primary crops there were dryland taro (*Colocasia esculenta*) and sweet potato. A detailed description of the mulching technique of dryland taro cultivation was given by Menzies (1920). Menzies also mentioned *ti* (*Cordyline fruticosa*) and sugar cane (*Saccharum officinarum*) on the rocky field boundaries.

The uppermost zone of the Kona field system was the *ʻamaʻu*, named for the native ferns (*Sadleria* spp.) that grew there (Kelly 1983). This area was still partially forested and was planted with bananas (*Musa* x *paradisiaca*). Other activities of the Hawaiians in this zone included felling trees for canoes, catching birds for feathered apparel, and collecting *māmaki* (*Pipturus albidus)* and other wild plants (Kelly 1983). Early explorers such as Captain King in 1778-79 estimated that the edge of the "wood" above Kealakekua was 10 to 11 km (6-7 mi) from the shore, although parts of the forest separated by cultivated lands reached to within 3 km (2 mi) of the coast (Beaglehole 1967).

Kohala Field System. Although the Kona field system may have been the largest and most developed on Hawaiʻi Island, the Kohala field system has been more intensively studied (Kirch 1985a). At its greatest extent, the Kohala system was 3 km (2 mi) wide and stretched for 20 km (12 mi) along the western flank of the Kohala Mountains from ʻUpolu to Kawaihae-uka (Rosendahl 1974). By the time of European contact, the field system was conspicuous even when viewed from the sea: Menzies noted the land laid out into small fields and commented on the sign of "industrious cultivation" while sailing to Kawaihae from Hāmākua with Captain George Vancouver in 1793 (Menzies 1920).

The most extensive archaeological work on the Kohala field system has been carried out at Lapakahi, an *ahupuaʻa* (land division) 1 km (0.6 mi) wide and 7 km (4 mi) long. Lapakahi was settled around 1300 A.D., possibly by people from the windward Kohala valleys; its population was first centered in small coastal communities primarily engaged in fishing (Griffin et al. 1971). For 200 years, the population increased and the uplands were developed for agriculture. Until about 1500 A.D., agriculture was shifting cultivation using slash and burn techniques and long fallow periods. Between 1500 and 1800 A.D., agriculture expanded, intensified, and became permanent (Rosendahl

1974). At the peak of population and production, the settlement pattern of the ahupua'a was characterized by a coastal zone of fishing villages; an intermediate, largely unoccupied zone directly upslope; and the upland agricultural zone about 3 km (2 mi) from the coast, where annual rainfall was 410 mm (16 in.) (Griffin et al. 1971; Rosendahl 1974). The upland and coastal zones were connected by a series of trails, indicating a close association of the two inhabited areas (Griffin et al. 1971).

The primary crop of the narrow, rectangular Lapakahi fields (and thus the Kohala system) was the sweet potato or 'uala (Ipomoea batatas), but at least a dozen other food and fiber plants were cultivated there (Rosendahl 1974). Carbonized remains of some of these crop plants have been recovered from excavations at Lapakahi (Griffin et al. 1971). Although the population supported by agriculture (and fishing) was quite large in 1850, and habitations were numerous both in the uplands and along the coast, the agricultural system and economy collapsed in the second half of the 19th century, due to severe decline in the Hawaiian population, destruction of planted fields by cattle, and economic upheaval with the conversion to a cash economy. By 1900, Lapakahi was abandoned, its remaining inhabitants moved elsewhere, and the area was incorporated into cattle ranches (Griffin et al. 1971; Rosendahl 1974).

Waimea Field System. Another major field system of the Island occurred at Waimea (Waimea-Lālāmilo). Early European visitors to Waimea such as Menzies (1920), who passed through the area in 1793, commented on the "plantations" and apparent fertility of the region. Thirty years later, Ellis (1827) described four villages of Waimea surrounded by plantations and estimated the population as 1,100-1,200. Some of the Hawaiian farms of Waimea were still cultivated as late as 1935 (Handy and Handy 1972), and numerous old Hawaiian fields and waterworks are known from the region (Barrera and Kelly 1974).

The Waimea agricultural system was located in an arc to the west and south of the present village and was composed of four separate complexes. One of the complexes is on the slopes above Waimea village, while the other three are on the plains near streams. Unlike the field systems of Kona and Kohala, the Waimea fields were irrigated, as indicated by the many remains of 'auwai or ditches, which diverted water from streams to the fields. However, this irrigation system was not the typical pond-field development such as is found in the windward valleys, but rather a "supplemental" system to allow periodic watering of fields or terraces (Clark 1983a). Only a few actual stone-constructed pondfields or lo'i for wetland taro cultivation have been found in the Waimea complex. In the upslope fields, ditches were constructed, apparently to drain the fields (Clark 1983a). The main crops were taro (Colocasia esculenta) and bananas (Musa x paradisiaca) (Clark 1983b). Stone mounds for the cultivation of sweet potatoes (Ipomoea batatas) and gourds (Lagenaria siceraria) have also been found (Clark 1983a).

The Waimea agricultural complex was occupied from the 13th or 14th century (Clark 1983b). At lower elevations between Kawaihae and Waimea, archaeological remains (mounds, cairns, and terraces) of dryland cultivation of sweet potatoes and gourds have been dated from 1600 to 1800 A.D. (Welch 1983). That Kawaihae was an important Hawaiian village is indicated by the immense Pu'u Koholā Heiau built nearby (Barrera and Kelly 1974). In 1824, Ellis (1827) estimated that Kawaihae had a population of 500.

In a discussion of early historic (1792-1850) vegetation patterns for the Kawaihae/ Waimea area, McEldowney (1983) recognized 12 vegetation types. The dry lowlands upslope of Kawaihae were called *pili* lands and were described by early European visitors as "barren" and treeless. These were grasslands, perhaps of the indigenous *pili (Heteropogon contortus)* and were probably maintained by Hawaiians through the use of fire. Above and to the north of the pili lands were the cultivated *kula* lands and a region called *ulu lāʻau*, in which fields and homes were scattered among native trees, probably *ʻōhiʻa (Metrosideros polymorpha)*. During this historic period and the previous period of Hawaiian occupation, native forests were still present on the Waimea plains and the slopes of Mauna Kea and the Kohala Mountains. McEldowney mapped four different mesic forest types, primarily of *koa (Acacia koa)* and *māmane (Sophora chrysophylla)*, to the south and east of the cultivated kula and ulu lāʻau lands. Additionally, an extensive band of ʻōhiʻa and ʻōhiʻa/koa rain forest was present upslope of the Hawaiian agricultural area, on the Kohala Mountains and stretching to the east above the Hāmākua coast. Clark (1983b) speculated that strong sociocultural pressures existed among the Hawaiians to prevent expansion of settlement and cultivation into the forested uplands and to maintain the forest resources for later exploitation.

Kaʻū Field System. Kaʻū is another district of the island of Hawaiʻi that supported a major Hawaiian field system; it is also the least well known. Newman (1972) reported two Kaʻū sites with extensive agriculture, one west and one north of Punaluʻu. The earliest European explorers, such as Captain King in 1779, were impressed with the barrenness of the many lava flows in Kaʻū and mentioned the scattered fishing villages along the coast; however, King wrote nothing of the area's agriculture (Beaglehole 1967). A lithograph made by an artist who accompanied Captain Vancouver in 1792-94 depicted an orderly, well-developed field system upslope of the village of Makakupu in Kaʻū, near the present Wood Valley (reproduced in Kirch 1985a). Forty years after Cook, Ellis (1827) noted that Kaʻū was populous near Waiʻohinu, Kaʻaluʻalu, and Kapāpala and observed that the population was scattered over the countryside where numerous taro *(Colocasia esculenta)* and sweet potato *(Ipomoea batatas)* fields were cultivated.

Waiʻōhinu was possibly the most important village of the district, because of its location on the only continuously flowing stream in Kaʻū (Kelly and Crozier 1972). Handy and Handy (1972) catalogued other important sites of habitation in Kaʻū; of these, Hīlea was perhaps the richest inland site for agriculture, supporting a number of crops, of which dryland taro was the most important. The true extent of the Hawaiian cultivation in these and other historical sites, as well as their possible connections to form an integrated Kaʻū field system similar to that of Kona and Waimea, is not understood, as the archaeology of these areas has not been thoroughly examined (Kirch 1985a). Newman (1972) speculated that a fifth field system may have been present in Puna, southwest of Kapoho, but much of this area was buried by the Kīlauea eruptions of 1955 and 1960, so the actual extent of past agriculture there may never be known.

Scattered Fields. While the bulk of the aboriginal Hawaiian population was supported by irrigated taro *(Colocasia esculenta)* cultivation in valleys or by the intensive farming of large field systems, scattered dryland fields in which shifting cultivation was practiced were also important in pre-contact Hawaiian agriculture. Dryland cultivation, particularly of taro and sweet potato *(Ipomoea batatas)*, was undertaken on all the Islands, even those where taro irrigation was widespread, such as Kauaʻi and Oʻahu (Handy and Handy 1972; Kirch 1985a). Areas that seem completely unsuitable for agriculture today were apparently cultivated by Hawaiians, perhaps during a pre-European contact population peak. Examples of seemingly non-arable sites are Kahoʻolawe, which supported sweet potato and yam (*Dioscorea* spp.) cultivation (Myhre 1970), and Barbers Point, Oʻahu, where archaeological remains suggest that between 1200 and 1870 A.D. Hawaiians were growing crops in mulched sinkhole pit gardens, behind windbreaks, and along margins of wetlands (Davis 1982).

On the island of Hawaiʻi, scattered fields interspersed with habitations were numerous along the lower windward slopes, where they were noted by many early European explorers and visitors. Captain King in 1778-79 noted that the slopes along the Hāmākua coast seemed "fully cultivated" (Beaglehole 1967). Forty years later, the naturalist James Macrae (Macrae 1972) and William Ellis (1827) observed many Hawaiian "plantations" in a band 6 to 10 km (4-6 mi) wide along the coast. For 64 km (40 mi) northwest of Hilo, forests were seen only above this zone of cultivation and on the bottoms and steep sides of ravines (Macrae 1972). McEldowney (1983) presented historical evidence for a band of agricultural land that stretched between Waipiʻo Valley and Hilo during the early historical period (until 1850). The zone contained actively cultivated fields, many fallow fields in which grasses and ferns grew, and groves of useful trees such as *hala (Pandanus tectorius)*, *kukui (Aleurites moluccana)*, mountain apple or *ʻōhiʻa ʻai (Syzygium malaccense)*, and breadfruit or *ʻulu (Artocarpus altilis)*. This Hāmākua coastal zone of habitation and cultivation was politically distinct from the nearby settlements of Waimea, from which it was separated by a dense rain forest crossed by foot-trails (McEldowney 1983).

While the Hilo area did not support a population as dense as that of leeward Hawaiʻi, during the early historical period Hawaiian settlements were found on Hilo Bay and in six villages along the Puna coast between Hilo and Cape Kumukahi, in areas with old lava flows and good soil development (McEldowney 1979). In marshy sites near streams and rivers of Hilo, wetland taro was cultivated (Handy and Handy 1972), supporting an estimated 400 people in the early 19th century (Ellis 1827). Many early visitors to windward Hawaiʻi who travelled between Hilo and Kīlauea mentioned two zones of cultivation: one near the shore 6 to 8 km (4-5 mi) wide and another more than 24 km (15 mi) above Hilo Bay, separated from the lower zone by dense forest (Bloxam 1925; Macrae 1972). The upland agricultural zone, with its upper limit near 460 m (1,500 ft) elevation (near present-day Mountain View), was characterized by small plots of taro and banana (*Musa* x *paradisiaca*), and much unwooded land covered by grasses, ferns, native shrubs, and scattered trees. These uncultivated lands were probably areas allowed to go fallow in the Hawaiian shifting system of slash and burn cultivation, but they also provided food in times of famine from arrowroot or *pia (Tacca leontopetaloides)*, *ti (Cordyline fruticosa)*, and the native *ʻamaʻu* fern (*Sadleria* spp.). Above this upland agricultural zone was the lower forest, where resources were periodically exploited by Hawaiians, but permanent cultivation did not take place (McEldowney 1979).

Puna was thought by Captain King in 1779 to be very sparsely populated (Beaglehole 1967). However, in 1823, Ellis (1827) reported relatively large populations in communities such as Kaimū, Kehena, Kamāʻili, Keaʻau, and ʻŌlaʻa, where people were cultivating taro, sweet potatoes, bananas, and sugar cane (*Saccharum officinarum*). Ellis commented on the groves of coconuts (*Cocos nucifera*) and *kou* trees (*Cordia subcordata*) at Kaimū and estimated the population at 725 with a total of 2,000 people in the vicinity. Coastal hala forest and "fern-covered plains" of Puna were modified and planted with taro; tree fern (*Cibotium* spp.) forests of the uplands were also used for taro cultivation after the ferns were uprooted (Handy and Handy 1972).

Cultivated lands of the Ka'ū District were either part of a field system or were scattered fields, drier but otherwise similar to those of Puna. Much of the Ka'ū and South Kona coast was too dry for agriculture but supported numerous small fishing villages (Ellis 1827; Beaglehole 1967).

Vegetation Replacement, Deforestation, and Erosion

Deforestation and erosion were the natural results of Hawaiian agriculture. There can be no doubt that Hawaiians greatly altered the lowland vegetation of the Hawaiian Islands, particularly during the period of expansion in population and intensification of agriculture between 1100 and 1650 A.D. (Kirch 1985a). Environmental changes associated with deforestation (apart from the simple loss of species) include increase in solar radiation; decrease in soil moisture, permeability, and surface water retention; faster run-off; lower water table and altered micro-climates; and drought (Newman 1969).

The forests of the irrigated valleys and windward slopes were often replaced completely by the taro ponds, gardens, habitations, and introduced tree species of the Hawaiians. Replacement was accomplished early in the sequence of Hawaiian cultural development by manually clearing the trees and burning vegetation. European botanists and naturalists who visited Hawai'i in the first few decades after Captain Cook's voyage to the Islands described the windward lowlands as a cultivated region several miles wide (Douglas 1914; Menzies 1920; Beaglehole 1967; Macrae 1972). Often such explorers encountered native plants only at the extreme heads of cultivated valleys or on ridges overlooking them. Erosion sometimes followed the clearing of valley slopes and nearby ridges for cultivation, and downslope effects on land fertility, water quality, runoff, and marine environments have been noted (Spriggs 1985). The capability of the land to grow crops, and undoubtedly the size of the human population, also influenced deforestation patterns and progress.

Ni'ihau and Kaua'i. In 1778, the island of Ni'ihau was described by David Samwell as a "low land entirely bare of trees" (Handy and Handy 1972). That this view was something of an oversimplification based on a short and limited visit is indicated by the several native tree and shrub species that St. John (1959) was able to collect on valley walls and cliffs of eastern Ni'ihau during two trips there in 1947 and 1949. These woody plants, as well as those collected on Ni'ihau by earlier botanists, are species adapted to dry forests and include *wiliwili (Erythrina sandwicensis)* and *kulu'i (Nototrichium sandwicense)*. One seemingly anomalous plant collected from Ni'ihau in the 1860s is *Delissea niihauensis*, thought by St. John (1959) to indicate the past occurrence of a moist forest in the uplands. However, Ni'ihau probably never had a rain forest, because of dry climatic conditions due to its position in the lee of Kaua'i (St. John 1982). Even though native woody plants did persist into the 20th century, it seems likely that the original dry and mesic forests of Ni'ihau were largely destroyed by Hawaiian land practices before the arrival of continental man.

The adjacent and much larger island of Kaua'i has fared far better than Ni'ihau and still retains a significant cover of native vegetation at higher elevations. Nonetheless, Hawaiian cultivation left its mark on that island, particularly in the densely populated, well-watered stream valleys of the northern and eastern sides. The valleys of the Nā Pali coast are archaeologically rich and display the remains of extensive irrigation systems (Kirch 1985a). Although undeveloped since their abandonment by Hawaiians, these valleys currently support native vegetation only at their extreme heads and on steep cliffs and ridges (Corn et al. 1979), indicating the replacement of natural vegetation by Hawaiian land practices. Of course, the presence of domestic cattle *(Bos taurus)* and feral goats *(Capra hircus)* after European arrival undoubtedly played a role in the displacement of native plants. The impacts of Hawaiian agriculture on many other lowland slopes and valleys of Kaua'i are difficult to assess, as most wet areas below 460 m (1,500 ft) elevation have been used for sugar plantations or cattle grazing after European contact (Armstrong 1983).

O'ahu. In a trip to Nu'uanu Valley, O'ahu, in 1825, Macrae (1972) estimated that he had walked 8 km (5 mi) through taro *(Colocasia esculenta)* fields and other crops before he reached the "woods." Even then, the first trees he mentioned were Polynesian introductions such as *kukui (Aleurites moluccana)* and mountain apple *(Syzygium malaccense)*, and it was only after Macrae continued inland that he began to see native trees such as *koa (Acacia koa)*, *kōpiko (Psychotria* sp.*)*, *'ōhi'a (Metrosideros polymorpha)*, and "lobelia" (probably a species of *Clermontia*). Meyen (1981) gave a similar description of his trip to Nu'uanu Valley and ascent to the *pali* (cliff) edge in 1831. It was only in the upper reaches of Nu'uanu, which had only a few huts and small cultivated fields, that Meyen saw the native 'ōhi'a. The density of the native vegetation and the variety of plant life impressed the young naturalist, and he listed dozens of native ferns, mosses, herbs, *Cyrtandra* species, and members of the family Lobeliaceae, which he observed and collected near the head of the Valley.

Apparently, Nu'uanu was fairly typical of southeastern O'ahu valleys, for Macrae (1972) commented that Mānoa Valley was also cultivated to its upper reaches. MacCaughey (1917) listed a few native plants that still occurred on the Mānoa Valley floor in the early 20th century and speculated that the lower talus slopes had been covered with native trees in Hawaiian times. In 1831 Meyen (1981) climbed to 245 m (800 ft) elevation above Punchbowl Crater before he began to record native plants. Kukui was prominent in his description of the vegetation, along with koa, tree ferns (*Cibotium* spp.), *olonā (Touchardia latifolia)*, and *Clermontia*.

Archaeological evidence of erosion has been gathered from several sites on O'ahu. Severe upslope erosion is indicated in Mākaha Valley, where irrigation systems were buried under material washed downstream (Yen et al. 1972). Degradation and erosion due to clearing and agriculture may have been responsible for the repeated flooding of Kamo'olāli'i Stream in upland Kāne'ohe during the period of Hawaiian habitation (Rosendahl 1976). The two most extreme examples of erosion during the prehistoric Hawaiian period are Kahana Valley, where an entire bay has been filled in by sediments from cleared lands upslope, and Kawainui Marsh, which was formerly on the coast (Kirch 1985a). Erosion from upslope forest clearing and swidden agriculture on adjacent slopes led to sedimentation and infilling of the Kawainui basin, which was first a bay, then a lagoon and fishpond, and later a marsh (Kelly and Clark 1980; Spriggs 1985). Some have speculated that erosion was encouraged by the Hawaiians to increase fertility of lowland agricultural areas.

Maui, Lāna'i, and Moloka'i. Although Maui is the second largest of the Hawaiian Islands, little archaeological work has been carried out there (Kirch 1985a). Many stream valleys and well-watered slopes of West Maui are known to have been settled early (Kirch 1974) and eventually supported large populations and intensive irrigation cultivation (Handy and Handy 1972; Kirch 1985a). Thus, vegetation in lowland sites of West Maui was probably replaced early during the Hawaiian period.

26

A similar replacement of natural vegetation by wetland cultivation of taro (*Colocasia esculenta*) probably also occurred in the lower valleys and slopes of windward East Maui, although the history and archaeology of this area are not well understood (Kirch 1985a). Archaeological investigations of leeward East Maui, however, indicate a large concentration of habitations and dryland cultivation between 400 and 700 m (1,310-2,300 ft) elevation in the district of Kahikinui. This complex of sites probably represents a field system similar to those of leeward Hawai'i Island and was apparently developed in the late prehistoric period as an expansion into a harsher, more marginal region (Chapman and Kirch 1979; Kirch 1985a). The Kahikinui area is part of the south slope of Haleakalā, where vegetation was recently surveyed by Medeiros et al. (1986). This study located many remnants of a rich dry-forest flora. A large Hawaiian population clearing land, setting fires, and gathering firewood in and near dry forests and shrublands could have severely impacted the native vegetation and contributed to its decline and present fragmentation. Likewise, a postulated agricultural area upslope of the coast between Kīhei and Mākena (Kirch 1985a) could have greatly disturbed natural dryland vegetation there, a remnant of which was described by Medeiros et al. (1984, 1986) on a very rocky substrate unsuitable for agriculture.

Even the high slopes of Haleakalā were visited by ancient Hawaiians, who used a shelter cave at 3,050 m (10,000 ft) elevation as early as the 9th century (Kirch 1974). Hawaiians were probably travelling to this area near the East Maui summit to gather adz material and to exploit the now-endangered *ua'u* or dark-rumped petrel (*Pterodroma phaeopygia*) for food (Kirch 1985a).

In Hālawa Valley on Moloka'i, evidence exists for erosion and slope instability resulting from shifting agriculture after clearing the *koa (Acacia koa)* and *'ōhi'a (Metrosideros polymorpha)* forest during an intensification of cultivation around 1200 A.D. (Kirch and Kelly 1975). Erosion of the upper valley and slopes over hundreds of years resulted in flooding, stream siltation, and lower valley infilling, producing land later cultivated during periods of agricultural intensification (Spriggs 1985).

Lāna'i and western Moloka'i were reported as treeless or "barren" by Captain Cook (Beaglehole 1967) and other European visitors. Menzies, on board Captain Vancouver's ship in 1793, observed no trees or shrubs on the south side of Lāna'i or the western end of Moloka'i and described the vegetation of these regions as "thin withered grass" (Menzies 1920). Thirty years later, Ellis (1827) also portrayed Lāna'i as "largely barren" but wrote that the people of Maui cut posts and rafters from the thickets of small trees found in gulches or ravines. It appears that these early accounts were not completely accurate: remnant dry and mesic forests exist today on Lāna'i in the upper reaches of gulches as well as other isolated localities (Spence and Montgomery 1976; Ziegler 1989). Ziegler (1989) estimated that half of Lāna'i could have supported dry forest or woodland before modification by Hawaiians and destruction by modern land use practices. While no native forests are extant on western Moloka'i, several native tree species were considered by Rock (1913) to be relatively common on Mauna Loa (on Moloka'i) in the early 1900s: *keahi (Nesoluma polynesicum), alahe'e (Canthium odoratum), 'ohe makai (Reynoldsia sandwicensis), kulu'i (Nototrichium sandwicense), maua (Xylosma hawaiiense)*, and the endangered gardenia or *nā'ū (Gardenia brighamii)*.

Kaho'olawe. The low, arid island of Kaho'olawe was likewise characterized by the early botanists and explorers as treeless with a vegetation cover of "coarse grass" and few shrubs, even before the advent of feral or domestic ungulates (Ellis 1827; Menzies

1920; Beaglehole 1967). During the mid to late 1800s, botanists who actually landed on the Island collected or observed a few native tree and shrub species but did not find extensive native plant communities (Myhre 1970). A few endemic *wiliwili (Erythrina sandwicensis)* trees still persisted on Kahoʻolawe in the 1980s (Corn et al. 1980). Kahoʻolawe was uninhabited or only sparsely inhabited when Europeans arrived in Hawaiʻi (Beaglehole 1967), but its eastern inland plateau apparently supported permanent settlements and dryland agriculture, particularly sweet potato (*Ipomoea batatas*) cultivation, during the period between the 15th and 17th centuries (Hommon 1980). Although severely impacted by introduced animals and later human manipulation, Kahoʻolawe presents one of the clearest pictures of extreme environmental degradation during the pre-European Hawaiian occupation of the Island. Hommon (1980) speculated that an increasing human population and the excessive clearing and burning needed for intensified agriculture resulted in "island-wide degradation," severe erosion (evidenced by archaeological remains buried under massive colluvial deposits), and the eventual collapse of the island's culture.

Charcoal fragments from Kahoʻolawe firepits and earth ovens (*imu*) were identified as native tree and shrub species characteristic of dry and mesic communities, none of which remain on the island today: *ʻāweoweo (Chenopodium oahuense)*, sandalwood (*Santalum* sp.), *kuluʻi (Nototrichium sandwicense)*, *koʻokoʻolau (Bidens* sp.), *alaheʻe (Canthium odoratum)*, and *ʻaiea (Nothocestrum* sp.) (Kirch 1985a). These findings indicate that much of the native woody vegetation of Kahoʻolawe was exploited for firewood, although it is possible that some of the wood could have been salvaged from driftwood.

Hawaiʻi. John Ledyard, who visited the island of Hawaiʻi in 1779 on Captain Cook's last voyage, noted that the woods "surrounded this island at a uniform distance of four or five miles from the shore" (MacCaughey 1918). While Hawaiian agricultural practices probably impacted lowland vegetation on both wet and dry sides of the island, the native vegetation of the large field systems of leeward Hawaiʻi was particularly affected by use of the land to grow crops; the development of agriculture in North Kohala led to the total replacement of the original dry forest and scrub of the cultivated area (Rosendahl 1972; Kirch 1985a). From a Kohala bog near Hawaiian settlements, Juvik and Lawrence (1982) provided evidence from pollen to substantiate the modification of native vegetation during the Hawaiian period. Their pollen analysis indicated a decrease in trees *(Metrosideros* and *Cheirodendron)* and an increase in herbaceous species over the last 1,600 years. By contrast, analysis of a core from a more isolated bog near the Saddle Road revealed little change in the amount of *Metrosideros* pollen over a period of 2,400 years. In an analysis of pollen from upland bogs on several other islands, Selling (1948) noted a retreat of rain forest and replacement by drier vegetation types during a period corresponding with the time of Polynesian occupation.

In the upper Kona and Waimea field systems, replacement of native vegetation seems to have been less absolute. The upper reaches of these field systems were described as being small, cultivated patches among native trees (Menzies 1920; McEldowney 1983). In 1794, Archibald Menzies (1920) observed native *ʻōhiʻa (Metrosideros polymorpha)* trees growing in the second zone of cultivation (*kaluʻulu*) at Kona, which was dominated by breadfruit or *ʻulu (Artocarpus altilis)*. David Nelson, the botanist who accompanied Captain Cook in 1778-79, was able to rather casually collect a number of native plants along the coast at Kealakekua and in the lowland cultivated zone up to 365 m (1,200 ft) elevation, among them several species that are today extinct, endangered, or extremely rare: *Scaevola coriacea, Achyranthes nelsonii, Solanum nelsoni, Neraudia cookii,* and *Kokia drynarioides* (St. John 1976, 1979). There is, however, evidence for erosion from strong winds after fields were abandoned in the Waimea area (Clark 1983b); wild cattle *(Bos taurus)* introduced later may also have played a role in degradation.

Even in areas with only scattered fields or shifting cultivation during the period before European contact, deforestation was widespread and often extreme. McEldowney (1979, 1983) catalogued the reports of early European visitors to windward Hawai'i, in which a band of unforested land along the Hāmākua coast and open, disturbed vegetation above Hilo were described. McEldowney believed these grass- and fern-covered slopes to be fallow fields of long-term swiddening or shifting cultivation, which were periodically burned by the Hawaiians.

Deforestation and changes in vegetation resulting from the Hawaiians' need for firewood have been recognized from several sites on Hawai'i Island. Kelly (1983) speculated that any lowland forest remaining along the Kona coast of Hawai'i would, by the 19th century, have been depleted by Hawaiians collecting firewood. She further theorized that the adaptation to the *imu* (earth oven) was an effort to conserve firewood and resulted in more "modest" needs of Hawaiians, compared with those people dependent on open fires. Based on current use in India, open fires might have required as much as two tons of firewood a year per family (Eckholm 1978). Identification of charcoal from firepits in or near the Waimea field system of Hawai'i Island gave results similar to those from the island of Kaho'olawe: all of the identified remains were of native tree and shrub species characteristic of dry or mesic forests (Murakami 1983). The most common type of charcoal recognized in Waimea belonged to the genus *Chenopodium* (probably *C. oahuense*), but other genera present were *Nothocestrum, Canthium, Nototrichium, Colubrina* or *Alphitonia, Diospyros, Sida,* and *Acacia.* The first four genera are the same as those found in Kaho'olawe firepits (Kirch 1985a), indicating a similarity in the original vegetation of the two areas that can hardly be imagined today.

Fire

Fire was the primary tool used by Hawaiians to clear lands prior to cultivation (Kirch 1982). This was true in areas adjacent to irrigated valleys and windward slopes as well as in the great field systems of the island of Hawai'i. Evidence for clearing and burning of vegetation has been found at many windward agricultural sites on several islands, such as Hālawa Valley, Moloka'i (Kirch and Kelly 1975) and Kamo'olāli'i Stream in upland Kāne'ohe, O'ahu (Rosendahl 1976). All three of the major Hawai'i Island field systems that have been closely investigated were found to have burn layers dated to the Hawaiian period, which in the lowest level contained land snail remains and charcoal indicative of the former vegetation. This is physical evidence of the Hawaiian use of fire to convert "vast tracts of native xerophytic and mesophytic plant communities" to cultivated fields (Kirch 1982).

A study of the Waimea field system on Hawai'i Island (Clark and Kirch 1983) included examination of fossil and subfossil land snails in a corridor from the coast to the upland field system (Christensen 1983). Abundant endemic land snail remains were found in a buried burn layer, while no native snails were found in the upper soil layers. The burn layer is interpreted to represent the interface between native forest vegetation and the cultivated fields and fired grasslands of the Hawaiian period of occupation. As might be expected, the land snail species of the lower Kawaihae/Waimea corridor were dry forest species, while those found in the upland Waimea area included extinct taxa thought to be wet and mesic forest inhabitants.

Botanical evidence for the composition of the pre-burn native forest of Waimea was derived from charcoal found in a burn layer from the Hawaiian era (Murakami 1983); these buried charcoals were identified as native trees such as *kauila (Colubrina* or *Alphitonia), lama (Diospyros sandwicensis), koaiʻa (Acacia koaia),* and shrubs such as *ʻāweoweo (Chenopodium* sp.*), ʻilima (Sida* sp.*),* and *ʻakoko (Chamaesyce* sp.*).* In an analysis of carbonized seeds of the Waimea burn layer, Allen (1983) identified several agricultural weed species and native shrubs that may have represented secondary succession on fallow or abandoned fields.

The leeward slopes of the Kohala Mountains that later became the Kohala field system were first occupied around 1300 A.D., and in the following 200 years the land was cleared and planted using slash and burn techniques (Rosendahl 1974). Peat cores from a Kohala bog at 1,100 m (3,600 ft) elevation were found to contain charcoal throughout the 1,600 years represented by the core, including the time of Polynesian occupation, when pollen shifted from that of native tree species to that of herbaceous plants (Juvik and Lawrence 1982). Carbonized plant material from the late prehistoric period of Lapakahi was almost completely composed of introduced agricultural species such as coconut *(Cocos nucifera),* yam *(Dioscorea* spp.*),* and sweet potato *(Ipomoea batatas)* (Griffin et al. 1971). The original dry forests and shrublands of leeward Kohala were probably similar in composition to those remnants described from upslope Kawaihae by McEldowney (1983).

The Kona field system had probably been occupied for more than 1,000 years when it was first seen by Europeans in 1778 (Kelly 1983). While forest burning and clearing for cultivation may have been relatively restricted during the first few hundred years of Hawaiian occupation, by the 16th and 17th centuries deforestation and cultivation had increased with the population to the level of intensity seen by the first European and American visitors of the 18th and early 19th centuries.

The more scattered fields of windward Hawaiʻi and other islands were carved out of the lowland forest by tree cutting and the use of fire. This shifting agriculture practiced on windward tablelands and less favorable leeward areas resulted in "even greater impacts upon the Hawaiian ecosystem" than did the irrigation agriculture of the valleys (Kirch 1982). Handy and Handy (1972) described methods of planting taro *(Colocasia esculenta)* in the *hala (Pandanus tectorius)* forests of Puna, in which some hala trees were felled to admit light, and cut hala branches were laid over the mulched taro and then burned. The resultant ash provided fertilizer, a fact appreciated by both ancient and modern practitioners of slash and burn cultivation worldwide (Bartlett 1962). Another type of vegetation the Hawaiians burned before cultivating was the wet forest of young substrates that had a dense cover of the matted fern, *uluhe (Dicranopteris* spp.*)* (Handy and Handy 1972).

Fire may have been repeatedly used to periodically clear the secondary growth on fallow fields. Many of the early European visitors to the island of Hawaiʻi described uncultivated but unforested lands of the lower slopes of Hāmākua (Ellis 1827; Menzies 1920; Macrae 1972). These were probably fallow fields of the shifting agriculture regime. In 1824, Bloxam (1925) noted a region more than 24 km (15 mi) above Hilo that had been recently burned by Hawaiians. McEldowney (1979) speculated that firing may have been done to encourage the growth of arrowroot or *pia (Tacca leontopetaloides)*, morning glory *(Ipomoea* spp.), and other types of plants used as "famine foods" and pig feed.

Apart from the use of fire before and after the cultivation of food plants, Hawaiians used fire to stimulate the growth of desirable plants in the leeward lowlands, particularly *pili* grass *(Heteropogon contortus)*. Pili, probably indigenous to the Hawaiian Islands (St. John 1973; Wagner et al., in press), was the preferred grass for thatching houses (Funk 1978) and was the primary material available in dry areas where alternate species such as *'uki (Machaerina* spp.*), ti (Cordyline fruticosa), loulu* palm *(Pritchardia* spp.), and hala were not found (Handy and Handy 1972). Some botanists have speculated that pili was a Polynesian introduction (Fosberg 1972); the grass is widespread in Polynesia and also occurs in India. Not surprisingly, pollen of this lowland grass has not been recovered from upland bogs (Selling 1948). Menzies (1920) in 1793 observed the Hawaiians burning a lowland grassland at Waimea, Kaua'i, to encourage new growth to use for thatch. While he did not identify the "rank grass" burned, it was probably pili. This grass was treated almost like a crop in the Kona District and lent its name to the lower zone of the field system there (Kelly 1983) as well as in the Kawaihae/Waimea area (McEldowney 1983).

Pili is adapted to survive fires (as are recently introduced grasses from Africa and North America), and it benefits from annual burning, which destroys many woody species (Vogl 1969). Few other native grasses have significant cover in the lowlands of Hawai'i; in most areas the natural lowland vegetation apparently lacked the fine fuels necessary for fire to carry with any great frequency. Most native Hawaiian plants are not adapted to survive frequent or intense fires, indicating that fire was not a significant ecological factor in pre-Polynesian plant communities (Mueller-Dombois 1981a). Exceptions are *koa (Acacia koa)*, a few other species of the montane koa parkland ecosystem, and several lowland shrub species such as *'ūlei (Osteomeles anthyllidifolia)* and *'a'ali'i (Dodonaea viscosa)*. In Hawai'i, it is now generally recognized that most, if not all, of the "lowland grasslands were anthropogenic in origin" (Kirch 1982) and resulted from the periodic firing of the lowlands, which over time destroyed the native woody species. Areas capable of supporting forests, such as 'Ewa, O'ahu, which today are covered by savannas of alien *kiawe (Prosopis pallida)*, were sparsely vegetated with grasses when first viewed by Europeans in the late 18th century. This is evidence that the original native forests had been destroyed by Hawaiian land practices (Davis 1982).

That Hawaiians used fire as a tool to modify their environment is not surprising, as pre-technological peoples all over the world have done so since the early Pleistocene (Stewart 1962); slash and burn agriculture is still practiced into the 20th century (Bartlett 1962). Use of fire in agriculture and vegetation alteration has been documented for other Polynesians such as the Maoris of New Zealand and for Melanesians (Davidson 1979). The North American Indians also manipulated their surroundings through the use of fire and may even have been partly responsible for the maintenance of the vast grasslands of the central plains of North America (Pyne 1982).

Polynesian Introductions

Plants. Hawaiian agriculture and concomitant deforestation and burning were physical disturbances to the land and original vegetation of the Islands. The Polynesians were also responsible for introducing nonnative plants and animals. Nagata (1985)

31

listed 32 plant species believed to have been brought to Hawai'i during the Polynesian period. Most of these were either observed or collected by botanists and naturalists who were on board the European ships that visited Hawai'i in the late 18th and early 19th centuries. Notable among the Polynesian introductions are food crop plants such as taro (*Colocasia esculenta*), bananas (*Musa* x *paradisiaca*), sugar cane (*Saccharum officinarum*), yams (*Dioscorea* spp.), and sweet potatoes (*Ipomoea batatas*). Taro, the staple food of Hawaiians, was represented in Hawai'i by more than 300 named cultivars, more than have been recorded for any other part of the Pacific (Abbott 1977). The crop plants of the Hawaiians all have their origins in Asia except the sweet potato, which was a separate and possibly later introduction from South America (Yen 1974). Although the coconut palm (*Cocos nucifera*) has in the past been considered indigenous, it is now generally thought to have been intentionally introduced by Polynesians as an important fiber and food source. In addition to food plants, Hawaiians introduced many other useful plants: *kukui (Aleurites moluccana)* for its oily nuts that provided illumination; *wauke (Broussonetia papyrifera)* as a source of tapa cloth; *ti (Cordyline fruticosa)*, useful for its leaves and roots; gourds (*Lagenaria siceraria*), essential in a culture lacking pottery; bamboo (*Schizostachyum glaucifolium*), the stems of which were fashioned into many different implements; and *'ōlena (Curcuma longa)*, for its rhizomes, which yielded a dye. Some of the Polynesian introductions were apparently not strictly cultivated but were allowed to grow in a semi-wild state. Among these are plants such as *'ape (Alocasia macrorrhiza)* and *noni (Morinda citrifolia)*, both used as sources of famine food.

A few weedy plants previously thought to be post-European contact introductions were recently found among the collections or notes of David Nelson, the botanist who accompanied Captain Cook in 1778-79 (St. John 1978). Among these weeds are yellow sorrel or *'ihi (Oxalis corniculata)*, *kāmole (Ludwigia octivalvis)*, *Urena lobata*, hairy merremia (*Merremia aegyptia*), and indigo (*Indigofera suffruticosa*). Except for indigo, these plants are relatively non-invasive and were probably brought accidentally by Hawaiians, associated with their food plants, animals, or gear.

Almost all the plant species introduced by Polynesians may still be found growing wild in the lowlands, often in areas previously cultivated or disturbed by Hawaiian habitation. Some, like taro, persist only sparingly near the streams formerly used for irrigation. Only a few of the Polynesian introductions have become abundant in the lowlands; among these are kukui, mountain apple or *'ōhi'a ai (Syzygium malaccense)*, and ti. Today, mountain apple and ti may be found in the understories of otherwise native *'ōhi'a (Metrosideros polymorpha)* and *hala (Pandanus tectorius)* forests, where they may have displaced native tree and shrub species. Many of the forests in which these early introductions predominate are probably successional after Hawaiian cultivation. Kukui may also be found in areas formerly cultivated (Wagner et al., in press), but it is also the primary tree on many windward lower valley slopes that seem too steep for agriculture. Of all the Polynesian introductions, kukui seems to have had the greatest capability for invasion in the lowlands, where it probably displaced forests of *koa (Acacia koa)* or mixed tree species (Fosberg 1972). Kukui has been found to escape from plantings and become invasive when introduced to other Pacific islands such as Christmas Island (T. Stokes, pers. comm. 1984).

Animals. Polynesians intentionally introduced food animals such as the pig (*Sus scrofa*), the dog (*Canis familiaris*), and the chicken (*Gallus gallus*). They also brought with them, probably inadvertently, the Polynesian rat (*Rattus exulans*), four species of geckos (*Lepidodactylus lugubris*, *Gehyra mutilata*,

Hemiphyllodactylus typus, and *Hemidactylus garnoti*), three types of skinks (*Cryptobleparus*, *Lipinia*, and *Emoia*), and land snails (*Lemellaxis gracilis, Lamellidea oblonga*) (McKeown 1978; Gagné and Christensen 1985).

The Polynesian pig was a domesticated form of the Eurasian wild boar, whose original range included most of Europe, the southern half of Asia, and many of the islands of Southeast Asia (Tisdell 1982). The Polynesian pig differed somewhat from today's feral descendents of domestic pigs in having a longer snout, underwool, and a straight tail (Stone, in press) and was apparently smaller than modern feral or domestic pigs. Captain Cook commented on the small size of Hawaiian pigs in 1779, estimating that the largest obtainable for his ships weighed only 23 to 27 kg (50-60 lb) (Giffin 1978).

It is thought that the Polynesian pig readily remained in domestication (Tomich 1986). Handy and Handy (1972) reported that only young pigs were allowed to run loose; older ones were kept in pens. David Douglas (1914) in 1834 told of a Hawaiian pig so tame that it came when called by its owner, who killed it and cooked it for his visitors. While early European visitors such as Cook (Beaglehole 1967), Ellis (1827), and Douglas (1914) all wrote of seeing domesticated pigs around the habitations of Hawaiians, botanists such as Macrae (1972) and Menzies (1920), who did considerable hiking in the Islands and wrote detailed journals, never mentioned seeing wild pigs in the native forests. Menzies described forests of the Kona slopes of Hawai'i Island that were so dense and filled with ferns and undergrowth that he was unable to walk through them, suggesting an absence of feral pigs, which typically damage undergrowth and open up the forest floor.

Polynesian pigs may have adversely affected areas near Hawaiian settlements and lowland forests in which they were allowed to roam, but their importance as food to the Hawaiians and careful husbandry probably ensured that destructive population levels would not be achieved. A more significant factor in vegetation alteration was likely the use of fire by Hawaiians on lower forested slopes to encourage the growth of plants such as arrowroot (*Tacca leontopetaloides*) that were used for pig food (McEldowney 1979).

The dog was a favored food animal of the Hawaiians and was also sometimes treated as a pet (Handy and Handy 1972). No mention is made by early European explorers of these animals running wild. They appear to have been kept in close domestication (Hommon 1976), where they were fattened on poi and other foods and were killed in large numbers for special occasions.

The third food animal introduced by Hawaiians, the chicken (or *moa*) was apparently allowed to become wild and occurred near houses and also in the woods (Handy and Handy 1972). Berger (1972) speculated that this bird would have ranged from sea level to 2,130 m (7,000 ft) elevation. Living on a diet of seeds and fruits, chickens may have competed with some native ground-dwelling birds, but they are unlikely to have had a serious detrimental effect on native plants.

The Polynesian rat was originally a native of Southeast Asia but spread to many Pacific Islands with the people who colonized them (Atkinson 1985); thus, interactions of the Polynesian rat with Hawaiian vegetation likely began about 1,500 years ago. The rat was probably a "stowaway" on Polynesian voyaging canoes (Kirch 1985a). Primarily a lowland species, where it is common in fields and forests (Tomich 1986), in recent years the Polynesian rat has been collected in Maui rain forests above 2,050 m (6,500 ft) elevation (Stone et al. 1984). The full impact of this Polynesian introduction may never be completely known, but Polynesian rats are especially likely to have negatively affected ground-dwelling birds of the lowlands (Kirch 1985a) and native invertebrates (Ramsay 1978; Gagné and Christensen 1985; F.G. Howarth, pers. comm.). The Polynesian rat may also have damaged native plants and altered community species composition through seed predation and girdling of soft-barked trees and shrubs, particularly during drought.

Remains of geckos, skinks, and nonnative land snails have been found during archaeological excavations of ancient Hawaiian sites of habitation (Davis 1982; Christensen 1983). Although lizards may have preyed on native insects of the lowlands, their true impact cannot easily be assessed (Gagné and Christensen 1985). The adventive land snails, which presumably were introduced with food propagules brought by the Polynesians (Kirch 1985a), are thought to have had little impact on native organisms (Gagné and Christensen 1985).

Use of Native Plants

The original Polynesian colonizers of Hawai'i found few native food plants to sustain them in the coastal and lowland regions they first settled (Handy and Handy 1972). However, as their culture developed, the Hawaiians found many native plants to be useful for medicine, clothing, utensils, and building materials; and a few were found to be edible. Apart from seaweeds or *limu*, of which more than 70 species were regularly used for food (Krauss, n.d.), most other wild food plants were important mainly in times of famine. Chief among these were ferns, particularly the tree fern or *hāpu'u (Cibotium* spp.) and *'ama'u (Sadleria* spp.). The starchy cores of fern trunks were cooked and then eaten by humans or fed to pigs (Handy and Handy 1972; Krauss, n.d.). As these fern species are important components of many native plant communities, heavy use for food could have greatly changed lowland wet forests (Kirch 1985a). Other ferns, such as *hō'i'o (Diplazium sandwichianum)*, *kikawaiō (Christella cyatheoides)*, and *pala (Marattia douglasii)*, were used for their edible fronds and rhizomes. The fruit of other native plants provided food, particularly shrubs such as *'ōhelo (Vaccinium* spp.*)*, *'akala (Rubus hawaiiensis)*, and *'ūlei (Osteomeles anthyllidifolia)*, and trees such as *lama (Diospyros sandwicensis)* and *loulu* palm (*Pritchardia* spp.). The pulp of *hala* fruits (*Pandanus tectorius*) was also eaten (Handy and Handy 1972). Apart from the loulu, which is today scarce in the lowlands, these native plants are still rather common and were probably not severely impacted by Hawaiian use for food.

A number of native plants known to have been used for medicine by the Hawaiians are listed by Handy and Handy (1972) and MacBride (1975). Many medicines were made from portions of plants such as leaves, fruit, or sap and could be collected without destruction of the target plant. Quantities of medicinal plants required were probably small. Many other native plants were good sources of fiber for the production of tapa cloth, baskets, mats, and cordage (Funk 1978). Most native fiber plants remain relatively common in existing native lowland and mid-elevation forests, including *māmaki (Pipturus albidus)*, used to make tapa; *olonā (Touchardia latifolia)*, *ōpuhe (Urera* spp.*)*, and *'ākia (Wikstroemia* spp.*)*, which were sources of cordage; and *'ie'ie (Freycinetia arborea)*, an important material for baskets and fish traps. Apparently some of these native fiber plants were maintained and partially cultivated in small forest plots (McEldowney 1979). Native plants were also used as dye sources; for example, the fruit of the *nā'ū (Gardenia remyi)* and the bark of *hōlei (Ochrosia* spp.), both uncommon trees, were used to produce a yellow dye

(Rock 1913; Neal 1965). *Koki'o (Kokia drynarioides)*, an endangered tree of leeward Hawai'i Island, may have been depleted by Hawaiians who collected its bark to produce a dye for fish nets (Rock 1913).

Building was also a consumptive use of native plants. Undoubtedly, the lowland forests near Hawaiian habitations were heavily used for house frames. Trees regularly used as timber included *'ōhi'a (Metrosideros polymorpha)*, *lama*, *naio (Myoporum sandwicense)*, and *kōlea (Myrsine spp.)* (Handy and Handy 1972). Canoe construction required the logging of large trees, particularly *koa (Acacia koa)*. In 1793, the naturalist Archibald Menzies (1920) described the felling of koa for canoes in Mauna Loa forests approximately 16 km (10 mi) upslope from Kealakekua Bay on Hawai'i Island. Menzies noted that the lower woods along his path had been considerably "thinned" of large koa trees by this Hawaiian practice. Many other Hawaiian woody plants were used for various utensils such as bowls (koa), spears (*alahe'e*, *Canthium odoratum*; *kauila*, *Alphitonia ponderosa* and *Colubrina oppositifolia*), and fishing gear (*hau*, *Hibiscus tiliaceus*; *wiliwili*, *Erythrina sandwicensis*). For the most part, common woody plants were used for everyday objects; one exception is the endangered *uhiuhi (Caesalpinia kavaiense)*, which possesses hard wood favored for use as *hōlua* sled runners (Handy and Handy 1972).

EARLY POST-EUROPEAN CONTACT PERIOD

Trade in Agricultural Products and Forest Resources

During the first few decades following the arrival of European ships in Hawai'i, the people of the Islands continued to practice subsistence agriculture. Beginning with Captain Cook, ship crews bartered with Hawaiians and exchanged iron and manufactured items for foodstuffs, trade that may have been considered gift-giving by the Hawaiians (Morgan 1983). As an increasing number of European and American ships began to visit Hawai'i to replenish their stores, regular trade in agricultural products began to develop (McEldowney 1979). The earliest ships traded for the staples of Hawaiian agriculture, primarily pigs (*Sus scrofa*), bananas (*Musa* x *paradisiaca*), taro (*Colocasia esculenta*), and sweet potatoes (*Ipomoea batatas*) (Menzies 1920; Beaglehole 1967). The number of crops in the Hawaiian Islands increased with the continual introduction of new cultivated plants by crews of sailing vessels and new residents. Nagata (1985) catalogued more than 100 species, primarily food plants, that were introduced in the 60 years after initial European contact. By the 1830s, a great variety of fruits and vegetables was available in Honolulu markets to provision ships (Meyen 1981).

Trade in foodstuffs and materials for caulking and ropes expanded with shipping traffic to the Hawaiian Islands. Fiber from the endemic *olonā* (*Touchardia latifolia*) produced extremely strong cordage valued by western sailors for ship rigging (Neal 1965). In the first half of the 19th century, the sandalwood trade, North Pacific fur trade, and whaling activities ensured that many ships would require provisioning in the Hawaiian Islands (Kuykendall 1938). Also during this period, crops (particularly sweet potatoes and white potatoes, *Solanum tuberosum*) began to be grown for export to the western United States (Barrera and Kelly 1974). The California Gold Rush of the 1840s created a demand for agricultural products that was partially met by Hawaiian farms; during this decade more than 80,000 barrels of potatoes were exported to the West Coast of North America each year (Creighton 1978). However, as the Hawaiian population had greatly decreased (Kelly 1983; Stannard 1989) and nonnative people were generally not yet allowed to own land (Kirch 1985a), export crops were probably cultivated on lands formerly used for Hawaiian subsistence and did not represent an expansion of agriculture. This changed after the mid-19th century, when the *ali'i* (ruling class) and foreigners began to develop large commercial farms to supply the "external market" such as whalers (Morgan 1983).

Firewood for Whaling Ships. The whaling industry contributed greatly to the early 19th century increase in trade at Hawaiian ports, particularly Lahaina, Hilo, and Honolulu. Whalers hunting right, bowhead, gray, humpback, and sperm whales (*Balaena glacialis, B. mysticetus, Eschrichtius robustus, Megaptera novaeangliae, Physeter macrocephalus*) in the North Pacific used Hawai'i as a provisioning stop and as a discharge point for whale oil and bone (Dorsett 1954; Culliney 1988). Whaling ships began to arrive in 1819, and more than 100 sailed to Hawai'i in each year of the 1820s. In 1822, two English missionaries counted 19 whaling ships in Kealakekua Bay and 24 in port at Honolulu (Culliney 1988). Whaling continued to accelerate, and by the 1840s the number of ships visiting Hawaiian ports each year exceeded 400 (Kuykendall 1938). For 30 years, whaling was the dominant economic activity of the Hawaiian Islands (Culliney 1988). This activity continued until the late 1860s, when poor whale catches, the introduction of petroleum for use in lamps, and the American Civil War effectively put a stop to the American whaling industry (Dorsett 1954; Kuykendall 1953). The amount of whale oil and bone shipped from the Hawaiian Islands greatly decreased after 1870 and was too insignificant to report after 1875 (Schmitt 1977). A local whaling industry concentrating on humpbacks in waters near Maui began in the 1840s and continued until humpback whales became rare in the late 1860s (Culliney 1988).

Whaling ships required large quantities of firewood to fuel the boilers that rendered whale blubber into oil. Firewood, like other provisions, was supplied in Hawai'i and may have had an "appreciable effect in reducing forest areas" (Kuykendall 1938). The trade in firewood to whalers was sustained for almost 50 years and must have resulted in the export of vast quantities of wood. In 1854, for example, 158 cords of wood (more than 20,000 ft^3, a volume representing at least several hundred trees) were supplied to ships at just one Hawaiian port, Hilo (Kuykendall 1938). Damage to lowland dry forests near leeward ports (e.g., Kawaihae, Lahaina) must have been even greater. In some years more than 500 whaling ships would call in the Hawaiian Islands (Kuykendall 1938), and a ship might return with as much as 1,000 barrels of rendered oil (Dorsett 1954). Demand for firewood came also from the populations of the growing port cities. On Kaua'i, the slopes above Kōloa were denuded in the 1840s and 1850s to provide firewood for ships, plantations, and export to Honolulu (Culliney 1988).

Sandalwood Trade. The trade in sandalwood or 'iliahi (*Santalum* spp.) had a greater impact on the Hawaiian environment than most other enterprises of the early 19th century. Although some sandalwood was collected and exported as early as 1791 (MacCaughey 1918), the period of intensive sandalwood export was relatively short lived, lasting only from 1815 to 1826 (St. John 1947). During this time, vast quantities of the fragrant heartwood were exported from Hawai'i to China, where it was used for chests, carved objects, and incense. The estimated amount of sandalwood exported from Hawai'i in 1821-22 was 1.8 million kg (4 million lb or 30,000 piculs) (Schmitt 1977). In the following season (1822-23), more than 1.1 million kg (2.5 million lb) of Hawaiian sandalwood were exported to Canton, China (St. John 1947). Bryan (1961a) estimated that more than 6,000 sandalwood trees would be required to fill the hold of one ship.

As a royal monopoly, sandalwood brought great wealth to the kings of Hawai'i, amounting to more than $400,000 in one year (Rock 1913). However, the Hawaiian royal family soon became indebted to traders, as they were at the same time paying large sums for western goods (Judd 1927). After the death of Kamehameha I in 1819, the collection of sandalwood accelerated to enable the Hawaiian kings to pay off the debt (St. John 1947). On a visit to the island of Hawai'i in 1823, Ellis (1827) reported that hundreds of Hawaiians of Waiākea were engaged in the search for sandalwood in the mountains of 'Ola'a and that villages of the Kohala District were nearly deserted for the same reason. Ellis also observed thousands of people engaged in hauling sandalwood logs to be shipped from the port of Kawaihae.

The result of this intense effort to collect sandalwood must have been a significant change in the species composition of the remaining low- and mid-elevation dry and mesic forests. While still of occasional occurrence in the middle of the 20th century, sandalwood was "decimated" on Oʻahu, particularly in the mountains behind Honolulu, and the low-elevation forests in which the tree was most common are now gone (St. John 1947). Contributing to the demise of lowland forest was the use of fire to detect sandalwood by the fragrant smoke produced when the tree burned. Frierson (1973) speculated that the logging and burning in the sandalwood era was a major factor in the destruction of the lower forests of Honouliuli, an *ahupuaʻa* (land division) on Oʻahu that was owned by the royal family. On the island of Hawaiʻi, sandalwood cutting is likewise thought to have contributed to the loss of forests between Kawaihae and Waimea (Barrera and Kelly 1974).

By 1831, when Meyen (1981) visited Oʻahu, the sandalwood trade was greatly reduced due to poor quality and depressed prices. When the U.S. Exploring Expedition under the command of Commodore Charles Wilkes explored the Islands in the 1840s, the scientists reported seeing only small shrubs of sandalwood on Oʻahu (St. John 1947). However, several hundred thousand pounds of sandalwood were exported annually between 1836 and 1839, perhaps from islands other than Oʻahu. Lesser amounts were shipped to China through 1876 (Schmitt 1977). Although sandalwood of saleable size and quality was eliminated from many lowland forests, particularly near centers of population, large sandalwood trees may still be seen in areas that were remote and less accessible in the 19th century.

Pulu. Another export trade detrimental to the Hawaiian forests was that of *pulu*, the soft reddish "hairs" that clothe the young fronds and stipe bases of Hawaiian tree ferns, particularly the *hāpuʻu-pulu* or *Cibotium glaucum*. This material was used to stuff pillows and mattresses. Pulu was collected by cutting off fern fronds and scraping the fibers off stipes (Doerr 1932), a process that often required the cutting or pushing over of large tree ferns (Hillebrand 1888). Between 1851 and 1884, several hundred thousand pounds were annually collected from the Kīlauea region of Hawaiʻi Island and shipped to North America (Neal 1965). Fifty to 75 people were employed in pulu collection and drying at the "Pulu Factory" near Nāpau Crater in what is now Hawaii Volcanoes National Park (Doerr 1932). Pulu was also collected in the Waimea region of Hawaiʻi until the tree fern had largely disappeared there (Barrera and Kelly 1974). Pulu collection and trading was an important source of income for the people of Kaʻū in 1860 (Kelly and Crozier 1972). The trade in pulu was flourishing as early as 1831, when Meyen (1981) noted that large quantities were gathered by Hawaiians to sell to foreigners. This enterprise continued and increased for several decades. Between 1860 and 1864, the annual export of pulu amounted to almost 272,000 kg (600,000 lb) (Kuykendall 1938), with more than 335,500 kg (738,000 lb) exported during the peak year of 1862 (Doerr 1932). After 1865, annual exports of pulu decreased, and the industry failed in the 1880s when superior stuffing materials replaced pulu, which had a tendency to mat, absorb moisture, and disintegrate (Doerr 1932; Schmitt 1977). A few decades later, for a brief period around 1920, the starchy cores of Hawaiian tree ferns were used for the commercial production of laundry and cooking starch (Krauss, n.d.).

Removal of the tree fern, a common understory dominant of lowland wet forests, may have seriously altered natural conditions, allowed the invasion of alien plants (Burton 1980), and even resulted in destruction of some native forests in over-exploited areas (Neal 1965). Hillebrand reported that pulu gatherers often "sacrificed the whole tree" to gather the pulu at the top, and in this way cleared away formerly "extensive thickets" of tree fern (Hillebrand 1888).

Wild Cattle, Goats, and Sheep

Cattle (*Bos taurus*) were first introduced into the Hawaiian Islands in 1793 and 1794 by Captain Vancouver, who landed animals at Kealakekua on the Kona coast of Hawai'i Island (Tomich 1986). A *kapu* (taboo) was placed on these animals to allow them to increase, and by 1802 cattle were numerous and destructive to Hawaiian farms of Waimea (Barrera and Kelly 1974). By 1823, there were "immense herds" of wild cattle in the Waimea area (Ellis 1827), and in the following decade great numbers of wild cattle were noted on Mauna Kea by explorers such as Macrae (1972) and Douglas (1914).

By the 1820s, John Parker and other resident foreigners had begun to hunt wild cattle in Waimea, first for salted beef to provision ships and later for tallow and hides destined for export (Wellmon 1969). The herds of wild cattle were decreased during the next 20 years as Parker and others began to capture and tame cattle to raise in ranching operations. Even so, wild cattle continued to be a problem on Hawai'i Island and on O'ahu, where such animals roamed near Honolulu and were rounded up and driven through the streets as late as 1850 (Kramer 1971).

In 1851, the island of Hawai'i was estimated to have 12,000 wild cattle and only 8,000 domestic animals (Henke 1929). By the mid-19th century, wild cattle were recognized as destroyers of Hawaiian forests and were blamed for converting the forests of Waimea into open plains, thus reducing moisture and precipitation (Anon. 1856). Wild cattle were still abundant in the *'ōhi'a (Metrosideros polymorpha)* forests between Waimea and the Hāmākua coast and at higher elevations on Mauna Kea when Isabella Bird visited the island in 1873 (Bird 1966). With the development of ranching, wild cattle were significantly reduced, although a large number, perhaps as great as 10,000, remained on Mauna Kea at the turn of the century (Hall 1904). Relatively small numbers of feral cattle, some recently feral from nearby ranches, continue to exist today on Hawai'i Island on the slopes of Mauna Kea, leeward Mauna Loa, and remote forests of the Puna District.

Goats (*Capra hircus*) and sheep (*Ovis aries*) were also introduced into Hawai'i in the late 18th century by European ship captains and established feral populations soon after their release (Tomich 1986). Like cattle, these ungulates began to impact natural vegetation almost 200 years ago and continue to degrade native systems in many areas. The effects of these and other feral animals are discussed in a subsequent section.

CHANGES IN VEGETATION SINCE 1850

Beginning in the 1850s, modern agriculture, ranching, and varied forest management practices, encouraged by efficient transportation and technologies, brought rapid and large-scale land use changes to Hawai'i. Rates and numbers of animal and plant introductions increased; fires became more important; and expanding human populations and expectations encouraged urbanization, industrial development, and an increasingly important visitor industry. Use of Hawai'i's natural resources, including native forests and watersheds, intensified.

Agriculture

While the beginnings of export agriculture and cattle production in Hawai'i date to the early 19th century, it was only after the "Great Mahele" of 1848 (see discussion in Kuykendall 1938:269-298) and subsequent land reform laws that large-scale commercial ventures were undertaken. These laws clarified land ownership, gave foreigners the right to own land, and allowed those Hawaiians who received land grants to later sell them to commercial interests (Kuykendall 1938; Creighton 1978). The land reform laws "provided the impetus for rampant clearing of upland forests" (Culliney 1988). The common people, who had previously farmed the land, were essentially dispossessed, and land ownership was concentrated in the hands of a few, first the Hawaiian chiefs but later non-Hawaiian planters and businessmen. Common Hawaiian people, who made up about 75% of the native population (Abbott 1977), received only 11,340 ha (28,000 a) of *kuleana* lands in the "Great Mahele" (Morgan 1983). By 1890, more than 25% of all private lands were owned by people of European descent, much of it purchased from the Hawaiian government in large parcels (Uyehara 1977).

Sugar. Sugar cane (*Saccharum officinarum*), the dominant agricultural product of present-day Hawai'i, was a major crop from the beginning of commercial agriculture in the Islands. Attempts were made to commercially produce sugar on O'ahu as early as 1819 (Morgan 1983). The first persisting sugar venture was Kōloa Plantation of Kaua'i, begun by American merchants in 1835 on land leased from Kamehameha III (Kuykendall 1938). Although the plantation founders went bankrupt, Kōloa Plantation continued to produce sugar into the 20th century (Vandercook 1939). More than 20 small sugar operations were active in the Islands in 1840 (Culliney 1988), and by 1846, there were 11 sugar plantations on Kaua'i, Maui, and Hawai'i (Kuykendall 1938). When resident non-Hawaiians became able to buy and lease land after the middle of the 19th century, the development of sugar plantations accelerated, although bad weather and unstable markets resulted in a drop in the number of plantations by 1857 (Culliney 1988). Isabella Bird noted that there were 35 sugar plantations in the Islands in 1873 (Bird 1966), up from the 29 that Mark Twain recorded on his 1866 visit (Morgan 1983). Bird described a visit to a sugar plantation at Onomea, Hawai'i Island, where fields stretched up to 460 m (1,500 ft) elevation and were bounded above by a dense, primarily native, forest of koa (*Acacia koa*), 'ōhi'a (*Metrosideros polymorpha*), and tree ferns (*Cibotium* spp.) (Bird 1966). Today, sugar cane fields still exist in the same area, and State Forest Reserve land occurs above the fields, but the forest has been greatly modified.

The early years of the sugar industry in Hawai'i were times of fluctuating demand and profit tied to economic booms and busts in North America. Following an 1876 reciprocity agreement reached with the U.S. government to allow open export of Hawaiian sugar into the country (Morgan 1983), sugar production increased, and the number of plantations rose to 63 by 1880 (Kuykendall 1938). Production increased approximately 20-fold between 1875 and 1898 to nearly 230,000 tons (Philipp 1953). In the last 25 years of the 19th century, there was a 10-fold increase in the amount of land used for sugar production, resulting in 50,610 ha (125,000 a) of cultivation (Kuykendall 1967). During this time of expansion, sugar plantations were extremely profitable, sometimes returning a 50% dividend on investment (Creighton 1978). In this era of profitability at the turn of the century, sugar plantations were found in a number of areas devoted to other uses today, such as the Kona region of Hawai'i Island (Kelly 1983) and Kāne'ohe, O'ahu (Rosendahl 1976). Plantations continued to be developed into the 20th century, particularly on the island of Hawai'i. The result was the conversion of much of the remaining lowland forest of the Hilo and Puna Districts into cane fields. As late as the 1890s, native rain forest still existed near Hilo along the road to Volcano, but in less than 30 years, most of the remaining lowland forest was cut and converted to sugar cane fields (Campbell 1920). Many converted lands of Puna were used to grow sugar for only a few decades, after which they became pastures, were allowed to revert to weed forests, or were subdivided.

By the decade of the 1970s, there were 15 sugar plantations in the Hawaiian Islands, with more than 101,210 ha (250,000 a) in cultivation; more than 1 million tons of sugar were produced in 1977 (Plasch 1981). In recent years sugar has become less profitable, and several plantations on O'ahu and Hawai'i have stopped production, a trend that is likely to continue as land becomes more valuable for residential purposes (Morgan 1983). By 1980, nearly half the Hawaiian sugar acreage was on Hawai'i Island, but both Maui and Kaua'i had nearly 20,240 ha (50,000 a) in sugar cultivation (Morgan 1983). Total area in cultivation dropped below 80,970 ha (200,000 a) in 1984 (Hawaii State Department of Planning and Economic Development 1985). In spite of acreage reductions, sugar production in 1988 remained fairly high at 979,000 tons, perhaps because of very high yields of more than 12 tons per acre (Hawaii Sugar Planters' Association 1988), far greater than yields obtained on U.S. Mainland sugar plantations (Morgan 1983).

During the period of expansion of the sugar industry, many native forests were cut to make way for new fields, and other forest areas were used for firewood to run boilers at sugar mills (Judd 1927). Culliney (1988) characterized the 1850s as the time when "the first great industrial eruptions in agriculture were about to spring up and roll over the major islands." Compared to the Hawaiian system of swidden agriculture formerly practiced on much of the lowland forest region, in which only small patches were cleared, the development of sugar plantations completely removed all native plant cover, and fields of sugar replaced vast tracts of forest. The lands that now support large-scale sugar cane cultivation are not the valleys and leeward slopes that were intensively cultivated by Hawaiians. Present-day plantations occur in areas, usually windward slopes, which previously were scattered fields and lowland wet forest (Newman 1972). After continuous sugar production for 150 years in many Hawaiian cane fields (Hawaii Sugar Planters' Association 1988), there is little likelihood of any native elements surviving.

When large acreages were cleared and planted to sugar, erosion followed, particularly on sloping lands. On O'ahu, erosion was actually promoted on slopes above the 'Ewa sugar plantation by agriculturists. Vertical plowing and drainage ditches encouraged erosion from the lower Wai'anae slopes and soil deposition on the lower plains (Frierson 1973).

Pineapple. The second-most important modern crop of the Hawaiian Islands is pineapple (*Ananas comosus*). This species was introduced into Hawai'i early in the 19th century (Nagata 1985) but did not become a commercial export crop until 1900 (Philipp 1953). Introduction of a sweeter pineapple variety and development of canning technology permitted the industry to become successful by 1901, when the Hawaiian Pineapple Company (later Dole Hawaii Division) was organized (Morgan 1983). Pineapple production rose rapidly during World War I (Culliney 1988). In the early 1920s, the entire island of Lāna'i was purchased by pineapple interests, and its upper plateau was devoted to pineapple cultivation. Pineapple plantations were also developed on Kaua'i, O'ahu, Maui, and Moloka'i, and by 1952 the area devoted to this crop amounted to 29,760 ha (73,500 a) (Philipp 1953). A small pineapple plantation existed for a short period near Kea'au on Hawai'i Island (McEldowney 1979), although pineapple was never an important crop on that island. Ten pineapple companies were in operation in Hawai'i in the late 1950s (Norbeck 1959); by the 1980s only three remained (Morgan 1983).

Pineapple cultivation remained profitable until the 1960s; before that time, 80% of the canned pineapple that entered the U.S. Mainland came from the Hawaiian Islands (Creighton 1978). From 1923 to 1947, annual shipments of canned pineapple to the mainland U.S. ranged from 65 to 295 million kg (144-649 million lb) (Schmitt 1977). By 1987, however, pineapple production had greatly decreased, and only 14,570 ha (36,000 a) of this crop were under cultivation (Hawaii State Department of Business and Economic Development 1988).

On the islands of Lāna'i and Moloka'i, much of the area developed for pineapple fields had previously been used for grazing cattle (Norbeck 1959; Hawaii State Department of Agriculture 1962); these areas were relatively arid lowlands, probably altered by the Hawaiian practice of burning during the pre-European contact era. All land used for pineapple on Lāna'i had not been severely degraded in the past; as late as the 1960s, native dry forest near Kānepu'u was cleared for pineapple cultivation (Ziegler 1989). Some pineapple fields on O'ahu were also old grazing lands (Frierson 1973), but other lands cleared for pineapple production had been covered by native *koa* (*Acacia koa*) and *'ōhi'a* (*Metrosideros polymorpha*) forests in the early 20th century (St. John 1947).

Coffee. Coffee (*Coffea arabica*), introduced before 1825 (Nagata 1985), was an early commercial crop in Hawai'i; more than 45,400 kg (100,000 lb) were being exported annually by 1852 (Kuykendall 1938). By 1902, the annual coffee export to the U.S. Mainland had increased to more than 454,500 kg (1 million lb). The amount exported grew steadily during subsequent decades, with peaks of more than 3.2 million kg (7 million lb) in the years 1932 and 1946 (Schmitt 1977). Although large-scale production is today restricted to the Kona District on the island of Hawai'i (Armstrong 1983), coffee was previously grown on O'ahu, Kaua'i, Maui, and other districts of Hawai'i Island (Philipp 1953). In the 1850s, Hanalei Valley on Kaua'i was "the coffee-growing center of the archipelago" (Culliney 1988). In the Kona District, coffee was planted for the most part on lands that were formerly part of the Hawaiian field systems, particularly in the zone of dryland taro cultivation. By 1898 more than 2,430 ha (6,000 a) in Kona supported coffee cultivation (Kelly 1983). In the Puna District of Hawai'i Island, coffee was grown for a brief period in the early 20th century on homesteads carved from the forests of 'Ōla'a (McEldowney 1979), and for a longer period in lower Puna. Judd (1927) deplored this destruction of large tracts of government-owned native forest, particularly since almost all the coffee homesteads soon failed, and some of the homesteaders subsequently sold their land to sugar plantations. The net result of "homesteading" was to convert government-owned native forests to privately-owned sugar plantations (Hall 1904). In addition to the direct loss of native forest by field clearing, some lowland mesic forests of Kaua'i and O'ahu have been degraded by the invasion of coffee plants, which in some areas dominate the understory and have probably displaced native plants.

Macadamia and Papaya. Two crops that have become important in the last few decades, particularly on the island of Hawai'i, are macadamia nuts (*Macadamia ternifolia*) and papaya (*Carica papaya*). A few hundred hectares of macadamia trees had been planted on several islands by the 1920s, and the area slowly increased to over 810 ha (2,000 a) in 1952 (Philipp 1953). By 1987 more than 8,500 ha (21,000 a) of land were devoted to macadamia nut production (Hawaii State Department of Business and Economic Development 1988). In 1980, almost 14 million kg (30 million lb) of macadamia nuts were produced in Hawai'i, and the crop had a value of more than $20 million (Scott and Marutani 1982). Most of the crop is grown on the island of Hawai'i, where 1,210 ha (3,000 a) of native forest near Kea'au were bought to be developed into a macadamia orchard in the 1950s (Philipp 1953). As macadamia nuts may be grown on rocky soils after bulldozing, areas formerly unsuited for agriculture may be cleared and used for this crop. In the district of South Kona, tall *'ōhi'a* (*Metrosideros polymorpha*) forests have been cleared to establish large macadamia orchards (Kelly 1983). Some cleared forests, such as those of the *ahupua'a* (land division) of Kapu'a on Hawai'i Island, were considered earlier this century to be among the richest lowland forests remaining in the Hawaiian Islands (Judd 1932; Rock 1913).

Although papayas were introduced as early as 1819 (Nagata 1985), this fruit became an export only in the 1930s. By 1978, more than 1,210 ha (3,000 a) were being used for papaya cultivation (Hawaii State Department of Planning and Economic Development 1985). Like macadamia nuts, papayas can be grown on very rocky soils after bulldozing. Many of the lowland areas now used for growing papayas, particularly in the Puna District of Hawai'i Island, had not been previously cultivated and supported forests containing native plants. Since papayas are often produced commercially for only two years before a field is abandoned (Muench et al. 1984), continued cultivation of this crop may be very land consumptive.

Other crops. Many other crops are grown in the Hawaiian Islands, particularly vegetables, fruits, and flowers (Philipp 1953). The production of flowers and foliage plants has greatly expanded in Hawai'i in recent years, and the ornamental plant industry now ranks fifth in Hawaiian agriculture (Yee and Gagné, in press); nursery products were valued at more than $30 million in 1982 (Hawaii Agricultural Reporting Service 1983). While economically valuable, the horticulture industry has also been responsible for the introduction of several hundred nonnative plants that have escaped and become naturalized in Hawai'i (Yee and Gagné, in press).

By 1930, more than 5,000 farms of less than 40 ha (100 a) existed in the Hawaiian Islands, many of them growing produce for consumption in nearby towns (Armstrong 1937). In 1979, 11,340 ha (28,000 a) of land in the State were devoted to diversified agriculture of vegetables and fruits, excluding pineapple, coffee, and macadamia nuts (Plasch 1981). This amounts to less than 10% of the total cropland in the Islands (Hawaii State Department of Planning and Economic Development 1985).

Bananas were an important export crop in the late 19th century. In 1889 more than 100,000 bunches were exported from Hawai'i. The amount exported fluctuated in subsequent years, but by 1922 more than 200,000 banana bunches were shipped from the Hawaiian Islands, with a value in excess of $200,000 (Pope 1926). As the population of Hawai'i increased in the 20th century, the banana crop was consumed locally, and by 1963, Hawai'i was an importer of bananas (Hawaii Agricultural Reporting Service 1983). While bananas continue to be grown commercially in Hawai'i, the local product cannot compete in price with imported fruit (Morgan 1983).

A number of crops have been tried commercially in Hawai'i and later abandoned. Rice (*Oryza sativa*) cultivation was important in the second half of the 19th century and the early 1900s. In 1887, Hawai'i exported about 6 million kg (13 million lb) of rice (Kuykendall 1967). Rice cultivation reached a peak of more than 3,640 ha (9,000 a) in 1909 but declined rapidly in subsequent years (Philipp 1953). Much of the land used for rice cultivation was in areas formerly irrigated by Hawaiians for taro production (Kuykendall 1953; Kelly and Clark 1980), but natural wetlands and streams were also used, resulting in great disturbance and "ecological changes" (Culliney 1988).

Corn (*Zea mays*) has also been grown on a large scale in Hawai'i; 4,050 ha (10,000 a) were planted in this crop in 1920 (Philipp 1953). By 1980, only 101 ha (250 a) of sweet corn were cultivated in the State, and most feed corn used by Hawai'i's livestock producers was imported (Hawaii Agricultural Reporting Service 1983). Today, some acreage in Hawai'i is used to produce hybrid corn seed. Even wheat (*Triticum aestivum*) was cultivated for a short period, particularly on Maui, and was exported from the Islands between 1855 and 1865 (Kuykendall 1953). Past commercial agricultural ventures in dry areas included tobacco (*Nicotiana tabaccum*), sisal (*Agave sisalana*), and cotton (*Gossypium barbadense*). All three were grown in the district of Kona on Hawai'i but were given up in the first half of this century (Kelly 1983). Sisal was also planted extensively on Kaua'i and O'ahu, where it has become naturalized (Degener 1938).

No discussion of agriculture in Hawai'i would be complete without mention of the illegal cultivation of marijuana or "*pakalōlō*" (*Cannabis sativa*). This crop may actually surpass sugar cane in monetary value (Morgan 1983). As cultivation is often carried out in State Forest Reserves, Natural Area Reserves, and other lands with native vegetation, there is great potential for damage to remaining forests (Anon. 1987). Medeiros et al. (1988) reported the introduction of 14 alien plant species into an upland Maui rain forest disturbed by marijuana cultivators. Of these "long-distance importations," eight persisted for more than a year, and half of these were deemed capable of spreading.

Koa and 'Ōhi'a Logging

For more than 1,000 years, the Hawaiians used island forests as sources of wood to build canoes, habitations, temples, and other structures, as well as sources of firewood. *Koa (Acacia koa)* was the preferred wood for Hawaiian canoes; Captain Cook reported 1,000 canoes at one time in Kealakekua Bay in 1779 (Beaglehole 1967; Apple 1971). In 1793, Menzies (1920) reported seeing numerous wood cutters and their trails in the forests above Kealakekua. While Hawaiians undoubtedly impacted the lowland forests and shrublands near their habitations, the lack of an extensive road system, draft animals, and metal tools limited the amount of logging they could accomplish and the area they could significantly affect.

With the influx of Europeans and Americans after European contact and the introduction of "modern" technology, both the demand for lumber and the ability to procure it increased greatly. Commercial logging of koa had begun by 1822 and became an established industry by the 1830s, primarily in the forests of the Kona and Hāmākua Districts on the island of Hawai'i (Jenkins 1983). In the 1830s, John Parker's ranching neighbors in Māna were also engaged in logging koa on the Mauna Kea slopes, hauling it to the coast, and shipping the roughly sawed lumber to Honolulu along with cattle (Wellmon 1969). These windward slopes of Mauna Kea, particularly between Makahālau and Hānaipo'e, were the site of the first major logging operation on the Island (Culliney 1988). This region supported dense koa forests in the 1850s but is today a grassland (Skolmen 1986a). In the 19th century, people living near koa forests, particularly ranchers and plantation owners, often built homes entirely of koa. Most of the native Hawaiians built their dwellings of the more common but difficult-to-work 'ōhi'a (Metrosideros polymorpha), using entire logs rather than sawing trees into planks and beams (Hall 1904).

The development of water- (and later steam-) powered sawmills accelerated the pace of logging on Hawai'i and Maui. Construction-grade lumber was shipped to Honolulu, but the availability of boards and timber imported from the U.S. Mainland and other Pacific islands kept prices down and depressed the local koa lumber industry (Jenkins 1983). By 1907, Hawai'i was importing 30 million board feet of construction lumber from the U.S. Mainland at a cost in excess of $100,000 (Nelson and Wheeler 1963).

Perhaps more important than commercial logging in the demise of the lowland koa forest on windward Hawai'i Island was the wholesale clearing of forested land for sugar plantations. Sometimes the forest was burned and the wood not even salvaged from the trees (Judd 1927). More often, wood-cutting and hauling were necessary on the plantation to provide firewood for workers (Bryan, n.d.) and to fuel the boilers for sugar processing (Judd 1927). As early as the 1860s, flumes extended miles upslope from plantations and ranches on the North Hilo coast, and after diverting streams to provide water flow, these flumes were used to transport wood downslope (Culliney 1988). In the 1920s, plantation camps on the Hilo coast used 8 to 10 railroad car loads of wood each week, derived primarily from Puna forests (Bryan 1961b).

Isabella Bird (1966), who traveled extensively on the island of Hawai'i in 1873, wrote of koa in lowland forests just above Hilo and commented on the numerous wood-hauling trails in the forest above an Onomea sugar plantation. By the 1880s, the extensive lowland forests of Hilo and Hāmākua (perhaps a nearly continuous band) that had once supported two sawmills were essentially gone, and the focus of the commercial logging industry moved to Kona and Maui (Jenkins 1983). By the end of that decade, most of the upper koa forests of Hāmākua had been converted into cattle ranches (Henke 1929).

On Maui, sawmills produced koa lumber at Ha'ikū, starting in 1860, and at Ka'ili'ili in the forests above Makawao, beginning in 1880 and continuing for 20 years. By the turn of the century, the Ka'ili'ili mill was reduced to providing firewood for the lowland sugar plantations (Jenkins 1983), and shortly afterward the accessible koa of Maui was gone. In the koa forests near Kula, clearcutting and burning had been going on since the 1840s (Culliney 1988). As early as 1873, the koa of the lower slopes of Haleakalā was severely depleted, evidenced by the desolation that Bird observed in the Makawao area (Bird 1966). Ironically, the literal meaning of Makawao is "forest beginning" (Pukui et al. 1974).

The absence of streams to run a sawmill kept the lumber industry of Kona on the island of Hawaiʻi at a more primitive level than that of Hāmākua or Maui, until steam-powered sawmills were established there in the 1920s (Jenkins 1983). However, it was these Kona forests that yielded the most sustained output of koa in the State. A small logging and milling operation was supported for more than 100 years, beginning about 1850. By this time, koa was used primarily for cabinet and furniture making and was no longer an important construction material. In a particularly self-contained arrangement, the Yee Hop Mill of Kona District supplied koa to Honolulu furniture makers, including a shop owned by C.Q. Yee Hop; land cleared of koa on the Kona slopes included the Yee Hop Ranch, which supplied meat to the Yee Hop Market of Honolulu (Jenkins 1983).

The use of logging to clear ranchlands was widespread during the 19th century development of ranches on the islands of Hawaiʻi and Maui. Parker Ranch and other large ranches expanded outward from original core areas, cutting trees, fencing, and stocking cattle. Not even lava-encircled forests of formerly protected *kīpuka* (islands of older vegetation) were spared (Culliney 1988). Much of the koa logging in the Kona District was carried out on forested ranchlands with the ultimate purpose of clearing the land for pastures. Logging and clearing has lasted for more than 50 years on one of the Greenwell Ranches of Kona (Jenkins 1983) and continues to the present day on privately-owned lands that support koa and ʻōhiʻa forests on Hawaiʻi Island. Warshauer and Jacobi (1982) documented the reduction of canopy cover over time through selective logging and expansion of grazing activities on an upland ranch of Mauna Loa. Their maps illustrate the progressive conversion of relatively intact, closed koa/ʻōhiʻa forest to one with only 25% canopy cover and disturbed understory, a process that started in the early 1950s and continued into the 1980s. In this area, logging was followed by bulldozing to speed the spread and establishment of nonnative grasses, and grazing operations expanded upslope following koa logging. Such ranch logging operations, often called "salvage," continue to supply koa wood to the local market. Some koa is currently shipped to Southeast Asia and California, where plywood is made; over 80% of the plywood is then sold to Hawaiian markets (E.M. Winkler, pers. comm. 1989).

In addition to koa logging, another early attempt at a commercial use of native Hawaiian forest trees was the export of railroad ties produced from ʻōhiʻa logged in Puna and South Kona Districts and milled near Pāhoa on the island of Hawaiʻi. In 1907, 13,000 ʻōhiʻa ties were shipped to the U.S. Mainland (Nelson and Wheeler 1963), and the Santa Fe Railroad contracted the Hawaiian Mahogany Lumber Company to supply 500,000 ʻōhiʻa ties annually over a period of five years (Jenkins 1983). The lumber company purchased logging rights to 8,100 ha (20,000 a) and leased 4,860 ha (12,000 a) of government forest to fulfill the railroad contract. ʻŌhiʻa railroad ties were exported until 1913. Nelson and Wheeler (1963) blamed the demise of this logging and milling operation on the fire that burned the Pāhoa mill in 1913. However, Jenkins (1983) reported that the Santa Fe Railroad did not renew its five-year contract because the ʻōhiʻa ties lasted only nine years in the climate of the southwestern United States, despite the rail company buyer's preliminary enthusiasm about the durability of ʻōhiʻa wood. During the first few decades of the 20th century, ʻōhiʻa from Puna was used commercially to produce flooring and wainscoting (Judd 1927). However, because ʻōhiʻa warps, checks, and is difficult to cure and work, its commerical use was supplanted by other woods when they became available (Bryan 1966).

Watersheds and Forestry

As early as 1846, the Hawaiian monarchs were concerned about the state of the Kingdom's forests, first passing a law making the forests government property and later, in 1876, legislating statutes to prevent forest destruction and resultant reduction in water supply (Lennox 1948). These attempts were clearly insufficient to halt the loss of forest cover due to continued development of agriculture and ranching and the spread and intensification of feral animals. Honolulu businessmen such as Charles Bishop and Sanford Dole were also cognizant of the great value of Hawaiian forests and were duly concerned about forest destruction by feral animals, insects, and wood cutting (Bishop and Dole, in Manning 1986). During the last year of the Hawaiian Kingdom (1893), the legislature created a new government department, the Bureau of Agriculture and Forestry. The Bureau employed a forester and an entomologist, operated a plant nursery, planted trees in Nu'uanu Valley and Tantalus on O'ahu, and distributed trees and advice on planting (Bryan 1961a). However, the Bureau's emphasis was on planting rather than on preservation of existing forest (Hall 1904).

Watersheds. After the loss of forests due to logging, clearing, ranching, and the depredations of grazing and browsing animals in the previous few decades, the 20th century brought an increased awareness of the importance of forest cover as watersheds for the sugar industry, so important to Hawai'i's economy. Hundreds of thousands of dollars were spent on water projects in the last three decades of the 19th century, primarily on development of ditches and artesian wells (Kuykendall 1967). By the 1920s, agricultural concerns had spent more than $20 million on irrigation systems and pumping plants (Judd 1927, 1931). The economic sense of developing such expensive systems is explained by the production statistics of 1895-1906, showing that irrigated plantations produced twice as much sugar as nonirrigated ones (Kuykendall 1967).

The reduction of native forest area was blamed for the diminished flow of water available for irrigation and fluming (Judd 1927). Concern was also voiced about the deteriorating state of O'ahu forests above Honolulu, which resulted in erosion, greater runoff, and a decrease in the amount of water available to recharge springs and artesian wells (Giffard 1918). At the turn of the century, a visiting U.S. Bureau of Forestry extension agent did not consider Hawai'i's native forests to be of commerical value (Hall 1904). Hawaiian forests were viewed as "protection forests" as opposed to "supply forests" (Judd 1927), and water was considered to be their most valuable product (Judd 1918, 1931; Hosmer 1959). Judd (1927) asserted that "water production will always be the paramount cause for the practice of forestry in these islands."

Forest Reserve System. Even prior to the establishment of territorial forest reserves, private landowners had set aside land to preserve as forests and had planted land with trees (primarily introduced species). At the beginning of the 20th century, large private forest reserves were maintained above sugar plantations at Līhu'e, Kaua'i (4,050 ha or 10,000 a) and Pāhala, Hawai'i (20,240 ha or 50,000 a) (Hall 1904). Early private tree-planting programs were carried out on Kaua'i, where 120 ha (300 a) were planted at Kilohana Crater in 1874 (Bryan 1961a), and a forester was employed by Līhu'e Plantation as early as 1881 (Bishop and Dole, in Manning 1986). H.P. Baldwin of Maui was cited as the "most extensive tree planter of the islands" for his planting of hundreds of thousands of trees, both native and alien species, on the lower slopes of Haleakalā (Hall 1904). In the 1890s, sugar interests of Hawai'i Island imported two foresters from Scotland to plant trees at Nā'ālehu in Ka'ū District and Kukuihaele in Hāmākua (Bryan 1961a). Also in 1890, the manager of Honouliuli Ranch on O'ahu, Harry von Holt, built fences and planted trees (primarily nonnative species) all over the southern Wai'anae Range to alleviate the treeless condition of formerly overstocked and eroded lands owned by Campbell Estate (Frierson 1973).

Despite these late 19th century private reforestation projects, Hall (1904) viewed the establishment of a government forest reserve system as absolutely essential to the preservation of Hawaiian forests (and thus watersheds). A Forest Reserve System became a reality in 1903, with the creation by the territorial legislature of a Division of Forestry to be directed by a Board of Agriculture and Forestry through a Superintendent of Forestry. The duties of the Division of Forestry as stipulated by Act 44 of the Legislature of 1903 were: "To devise ways and means of protecting, extending, increasing, and utilizing the forests and forest reserves, more particularly for protecting and developing the springs, streams, and sources of water supply, so as to increase and make such water supply available for use" (Bryan 1961a).

The first forest reserve was established in 1904 and included 370 ha (913 a) of O'ahu land owned by the Territory (Hosmer 1959). By 1906, the infant Forest Reserve System contained a total of 136,490 ha (337,140 a), two-thirds of this on the island of Hawai'i (Lennox 1948). The System grew steadily under the first Superintendent of Forestry, Ralph Sheldon Hosmer, who came to Hawai'i in 1904 from the U.S. Forest Service and remained until 1914. His primary accomplishments were the organization of the Division of Forestry and the establishment of 37 forest reserves with 323,890 ha (800,000 a) (Bryan 1961a). While forest reserves were important watersheds, their boundaries were drawn "so as not to interfere with revenue-producing lands," and such lands were not generally thought to be useful for agriculture (Judd 1918).

The expansion of the Forest Reserve System continued under Charles S. Judd, the second Superintendent of Forestry, who served in this position from 1915 until his death in 1939 (Hosmer 1959). By 1936, the territorial forest reserves encompassed more than 400,000 ha (1 million a), more than half of which were on the island of Hawai'i. Kaua'i, O'ahu, and Maui each contained nearly 60,730 ha (150,000 a) of forest reserve land, and Moloka'i had almost 20,240 ha (50,000 a) in the System (Lennox 1948). A few years earlier, Judd (1931) reported that 25% of the Territory's land area was in forest reserves. C.S. Judd was a progressive administrator for Forestry, who clearly was looking to the future when he wrote: "It requires no great imagination to see that at no distant time, by reason of the inevitable increase in population and increased development of every sort, we or our descendants shall be using a vastly greater amount of water than we do today" (Judd 1924).

Private lands were an important part of the early territorial Forest Reserve System, amounting to more than one-third of the total acreage in 1948 (Lennox 1948). During the 1930s, C.S. Judd worked for the aquisition by the Territory of private forest reserve lands, particularly those comprising the Honolulu watershed on O'ahu, but he was not entirely successful in this endeavor (Hosmer 1959). By 1952, the Forest Reserve System contained 64 reserves, but only 65% of the total area was owned by the Territory (Hosmer 1959). Nearly all of this reserve land was still considered to be of "primary value as watersheds" (Lennox 1948). Many private lands remained in the reserve system controlled by the Territory (and later State) until the 1960s, when changes in the laws regulating forest reserves caused many landowners to withdraw their forests from the System (Little and Skolmen 1989). By 1975, 98% of the private lands formerly placed in the State Forest Reserve System had been removed from that status (Warshauer 1986). The current State law regarding forest reserves (Hawaii Revised Statutes, Title 12, chapter 183-11 through 183-15) allows for the surrender of private lands to the government for forest or water reserves and provides exemption from taxes for such lands; however, the private landowner is required to surrender control of reserve lands for a term of more than 20 years.

There were three primary emphases of the territorial forestry program during C.S. Judd's 25-year tenure: remove wild stock, fence, and plant (Judd 1931; Hosmer 1959; Bryan 1961a). These three emphases are evident in the monthly progress reports that L.W. Bryan submitted as Assistant Territorial Forester on Hawai'i Island. Bryan carefully detailed the number of feral pigs, goats, sheep, and cattle *(Sus scrofa, Capra hircus, Ovis aries, Bos taurus)* killed in each reserve (on Hawai'i Island). From 1926 to 1927, the monthly total for Hawai'i Island reserves ranged from less than 50 to more than 1,100 animals removed (Bryan 1926-1927). Also reported were miles of fence newly laid and repaired, again according to forest reserve. Numbers of trees distributed and planted were accounted for, and the planting activities of private plantations were detailed as well.

Bryan summarized 25 years of forestry work (1921-46) on the island of Hawai'i in the *Hawaiian Planters' Record* (Bryan 1947). The amount of effort expended to protect forest reserve lands described in this summary is impressive. During the 25-year period, over 480 km (300 mi) of fencing were constructed on forest reserve boundaries, but the total fencing workload was nearly 800 km (500 mi) when fence rebuilding and repair were considered. In the same period, more than 0.25 million feral animals were removed from Hawai'i Island forest reserves and adjacent lands, resulting in an average annual kill of 10,000 animals. Feral animals removed were primarily goats, sheep, and pigs but also included cattle and donkeys, dogs, and horses *(Equus asinus, Canis familiaris, Equus caballus)* (Bryan 1947).

The figures cited by Bryan (1947) for tree-planting are, if anything, more impressive than those for fencing and feral animal control. On Hawai'i Island alone, more than 4 million trees were planted on forest reserves by territorial Forestry Division personnel with the help of the Civilian Conservation Corps. In addition to tree seedlings and saplings, 4,545 kg (5 tons) of tree seeds were planted on reserve lands. The nursery operation of the Hawaii District Forestry Division was also highly "productive," with six plant nurseries credited with propagating and distributing more than 8 million plants in the period from 1921 to 1946. Much of the watershed planting work was discontinued with the end of the Civilian Conservation Corps and the U.S. entry into World War II in 1941 (Hosmer 1959). Although an important part of Judd's forestry program, tree planting was not a new idea in Hawai'i; it is estimated that 12 million trees were planted in Hawai'i in the 140 years from 1778 to 1920 (Bryan 1961b).

Planting of Alien Trees. Plants selected for large-scale planting on forest reserves were generally not natives. Although more than 70 native tree and shrub species were outplanted in forest reserves in the Islands between 1910 and 1960, the recorded numbers of native trees planted pale to insignificance next to the vast plantings of nearly 1,000 alien species (Skolmen 1979). The most commonly planted native tree was koa *(Acacia koa)*, with more than 1 million plants placed in forest reserves on five islands over a period of 50 years.

In the early days of the territorial forest reserve planting program, native species and, in fact, any commercially valuable timber trees were considered unacceptable for planting and were avoided. Harold Lyon, head of the Department of Botany and Forestry of the Hawaii Sugar Planters' Association and an influential figure on the Board of Commissioners of Agriculture and Forestry, believed that the native forests were "doomed" so there was no point in planting native species (Lyon 1918). The phenomenon of *'ōhi'a (Metrosideros polymorpha)* dieback was perceived as a disease with "no practical remedy," and natural forest regeneration was thought to be unlikely in dieback areas (Bryan, n.d.). Since the importance of forest reserves was thought to be their capacity as watersheds, Lyon felt that it was necessary to introduce a "new flora," preferably of tree species that were fast-growing, to protect slopes denuded by feral animals and cattle grazing. Native trees were generally thought to be too slow growing to be useful in reforestation (Judd 1931). It was also believed impossible to grow native plants on the altered soils of disturbed native forests. C.S. Judd was apparently not in complete agreement with Lyon and others as to the reforestation potential of native species, for he reported the reforestation of two O'ahu reserves with koa and *kukui (Aleurites moluccana)*, the latter then thought to be indigenous (Judd 1918).

Most of the tropical trees introduced for their rapid growth were known to have little use for timber or other wood products. It was important that introduced tree species be of no value commercially, so that future generations would not be tempted to cut timber or fuel wood in these watershed forests (Bryan 1961b). Fast-growing trees, such as blue gum *(E. globulus)* and paperbark *(Melaleuca quinquenervia)*, were planted in forest reserves even though they were known to have no potential for timber. Swamp mahogany *(Eucalyptus robusta)*, the most commonly planted alien tree, was primarily selected because of its rapid growth but was also considered a good timber species (Nelson and Honda 1966; Skolmen 1979).

Lyon's candidate for the perfect tree for forest reserve plantings was *Ficus* or banyan (Lyon 1918). Banyans or figs were deemed desirable for watersheds because they would grow in a range of soils and climates, be spread by birds, establish in *uluhe (Dicranopteris)*-dominated areas, and start life epiphytically on dead trees (Culliney 1988). Lyon also thought that banyans, with their multiple trunks and tangled aerial roots, could be used to form a barrier to cattle on forest reserve boundaries (Lyon 1923). In their enthusiasm to carry out Lyon's plan to replace existing forests of Territorial reserves with *Ficus*, foresters of the Territory and the Hawaii Sugar Planters' Association spread banyan seed by hand on stumps and logs (Lyon 1923) and scattered it from airplanes, depositing more than 23 kg (50 lb) of *Ficus* seed over forest reserves on O'ahu in just one flight (Lyon 1949). A total of 37 species of *Ficus* was planted in the forest reserves of the Hawaiian Islands (Skolmen 1979); many were of undetermined identity and designated only by seed lot number (Bryan 1947). Although certain species of *Ficus* grew well and reproduced in some reserves (e.g., Hilo Forest Reserve) (Bryan 1961b), they were unsuccessful in other areas and were replaced by plantings of other alien species (Bryan 1977a).

Trees and shrubs from tropical areas all over the world were planted in Hawai'i's forest reserves. Although later plantings focused on swamp mahogany and other potentially useful timber species, many plants were introduced because they were ornamental, bore edible fruit, or grew quickly. Some introduced species turned out to be highly invasive and disruptive to natural communities. A case in point is firetree *(Myrica faya)*, which was planted in at least ten forest reserves on the islands of Kaua'i, O'ahu, and Hawai'i (Skolmen 1979). This fast-growing, nitrogen-fixing tree rapidly spread from plantings and was already considered a pest by 1944 (Neal 1965). Even though firetree was declared a noxious weed (Hawaii State Department of Agriculture

1962) and State agencies took steps to control it in pasturelands, eradication was not achieved, and the tree is an increasingly serious problem on both public and private lands. Other examples of beautiful but disruptive alien species planted in forest reserves since the 1920s are Christmasberry *(Schinus terebinthifolius)*, strawberry guava *(Psidium cattleianum)*, glorybush *(Tibouchina urvilleana)*, and shoebutton ardisia *(Ardisia humilis)* (Skolmen 1979).

Few controls were in place earlier this century to prevent the wholesale importation of alien plant species into Hawai'i (Smith 1989a). The correspondence over several decades between L.W. Bryan and H.L. Lyon contains many references to alien plants collected by both men during their travels. One example is a 1946 letter, in which Lyon reported his intentions of bringing seeds of *Macaranga, Melochia*, and "many trees" from Samoa to plant in Hawai'i (Lyon 1946). Bryan (1980) credited Lyon with introducing into Hawai'i the seeds of almost 10,000 species. Lyon also obtained seeds through correspondence with people in other countries and was instrumental in sending others on foreign collecting expeditions. L.W. Bryan was sent on one such trip to the Far East, where he collected more than 500 different plant taxa for introduction into Hawai'i (Bryan 1977a).

Commercial/Industrial Forestry. The emphasis on forest reserves as watersheds began to change during the administration of William Crosby, the third Superintendent of Forestry, who served in that capacity from 1939 to 1955. During Crosby's tenure, experimental planting of introduced trees accelerated, and large-scale plantings of timber species began in forest reserves (Hosmer 1959). By the 1960s, the concept of "multiple use" of forest resources was promoted by the State Forestry Division, with two or more combined uses seen as necessary because of Hawai'i's "land shortage" (Nelson and Wheeler 1963). One of the accepted multiple uses was forage production for cattle *(Bos taurus)* or feral game animals, a complete reversal from the conviction of earlier foresters that grazing and browsing animals must be excluded from forest reserves. After World War II, sport hunting of feral game animals became an important use of forest reserve lands, and sustained yield management of these animals began to be practiced. With time, many fences built in previous decades to exclude cattle began to deteriorate, and little new fencing was undertaken (Warshauer 1986, 1988). Judd (1918) may have anticipated such a policy reversal when he wrote, "To make appreciable reduction in any of the forest reserve areas on the plea of stock production or on other grounds, would be disastrous to the main purpose for which they were created."

The three decades following the close of World War II were a time of great change in Hawai'i, including statehood, increased urbanization, and population growth. This period saw the continued planting of timber species on forest reserves and the initiation of plans for large-scale industrial forestry in Hawai'i. While Hawai'i, particularly Hawai'i Island, had more than 400,000 ha (1 million a) of what was called "commercial forests," most of this land was covered with native 'ōhi'a *(Metrosideros polymorpha)* and was considered to be inadequately stocked for a commercial logging operation. 'Ōhi'a was judged to produce timber of poor quality, difficult to work when compared with alien species such as silk oak *(Grevillea robusta)* (Nelson and Wheeler 1963). 'Ōhi'a wood was also known to check and warp badly (Bryan 1966). Therefore, plans to exploit the State's forest resources in the two decades following 1960 generally expressed the need to replace 'ōhi'a forest with "more valuable" planted alien timber species (Nelson and Wheeler 1963; Hawaii State Department of Land and Natural Resources and Department of Planning and Economic Development 1976). Because of superior growth, stocking density, and uniform size, planted stands of nonnative trees (which constituted only 2% of the commercial forest area in the early 1970s) represented nearly 40% of the State's sawtimber (Hawaii State Department of Land and Natural Resources and Department of Planning and Economic Development 1976). Until 1960, the most commonly planted alien tree in Hawai'i's forest reserves was swamp mahogany *(Eucalyptus robusta)*, followed by silk oak and paperbark *(Melaleuca quinquenervia)*, the last being of no commercial value (Skolmen 1979). Large-scale planting was restricted to a few tree species of known value and performance, to ensure that commercial wood producers would have adequate raw materials for their operations (Bryan 1957). In the last few decades, much research work has been conducted by the U.S. Forest Service and the Hawaii Division of Forestry and Wildlife to assess the growth, performance, and wood characteristics of widely planted timber species, particularly *Eucalyptus*, silk oak, pines *(Pinus* spp.), and tropical ash *(Fraxinus uhdei)*, as well as trees planted on a more limited trial basis (Skolmen 1974, 1986b; Buck and Imoto 1982).

In the early 1960s, one forester speculated that timber could become the basis of an industry that would equal or surpass sugar in jobs provided and income produced in Hawai'i (Nelson 1963). In 1965, State and Federal foresters estimated that 40,490 ha (100,000 a) of forest land were needed for timber production just to supply the State's current demand for wood (Nelson and Honda 1966). To realize the State's goal of full-stocking on Hawai'i's commercial forest lands would require planting alien timber species on 200,430 ha (500,000 a), roughly all the commercial forest land owned by the State. Most of this area was 'ōhi'a and 'ōhi'a/*koa (Acacia koa)* forest (Nelson and Wheeler 1963). Only the Hāmākua and Hilo Districts of Hawai'i Island were considered to have enough plantings to support a lumber-milling operation (Nelson and Honda 1966). By 1975, forestry potentials were examined by the Hawaii State Department of Land and Natural Resources and Department of Planning and Economic Development (1976), with three alternative levels of timber production detailed: the "current" level of clearing and planting, which involved 12,150 ha (30,000 a) and use of existing plantations; an ambitious program requiring clearing 44,530 ha (110,000 a) of native forest to provide a "basic forest industry;" and an even more ambitious level of planting, which would require over 117,410 ha (290,000 a) of native forest and would support an "extensive industry."

The actual development of industrial forestry on State lands in Hawai'i has, so far, been much less than what was proposed in the 1960s and 1970s, but nonetheless, much native forest acreage has been converted to timber plantations. While the 1930s were a time of intensive tree planting in Hawai'i, with thousands of acres a year cleared and planted, the following two decades saw only a few hundred acres of forest reserves annually converted to plantings (Schmitt 1977). Bryan (1966) estimated that 1,010 ha (2,500 a) per year were being cleared and planted with timber species in the early 1960s. The State Division of Forestry reported that more than 3,640 ha (9,000 a) of forest reserve land were planted with timber between 1961 and 1966, followed by 3,240 ha (8,000 a) during the next five years. During both five-year increments, the actual acreage planted fell short of the State's goal (Hawaii State Department of Land and Natural Resources 1974).

Between 1960 and 1970, more than 4,450 ha (11,000 a) were cleared and planted for a single project in Waiākea and 'Ōla'a Forest Reserves on the island of Hawai'i. This one project represented more than half the State acreage planted with alien trees during the decade (Schmitt 1977). The Waiākea/'Ōla'a project was accompanied by controversy: the use of native forest for planting alien timber species was criticized and opposed by University of Hawaii professors and other scientists, who decried the loss of habitat for endangered species (Benson 1970; Anon. 1970). By contrast, the State Forester during this period was extremely critical of subsequent endangered species legislation and believed that "we must be permitted to modify and develop our native forest" (Tagawa 1976).

In 1973, an additional 2,150 ha (5,300 a) of native forest in the Waiākea and 'Ōla'a Forest Reserves were proposed for clearing and planting at a cost of more that $1.5 million (Hawaii State Department of Land and Natural Resources 1973). In the Division of Forestry's Environmental Impact Statement (EIS), the need to reduce the State's dependence on imported wood was cited; Hawai'i was producing only 2% of the lumber it used. The bulldozing of native forest to plant alien tree species was not acknowledged to be a negative impact: "Whether or not this is an adverse effect is a matter of personal viewpoint and judgement" (Hawaii State Department of Land and Natural Resources 1973). The Animal Species Advisory Commission, which reviewed the State's five-year planting plan and the EIS for the project, recommended that the plan be rejected for both environmental and economic reasons, and that "a moratorium be declared on clearing and planting operations for exotic timber production in areas of intact native vegetation" (Animal Species Advisory Commission 1974).

By the mid-1970s, very little planting was being done on State forest reserve land, amounting to only 40 to 150 ha (100-360 a) per year (Schmitt 1977). Actual plantings were only 7-30% of the area proposed by the State Division of Forestry (Hawaii Department of Land and Natural Resources 1974). Bryan (1977a) blamed this reduced planting effort on the lack of economic success of earlier programs. By 1984, the total area of planted forest in the State of Hawai'i was 18,840 ha (46,525a), almost half of it on the island of Hawai'i. For comparison, the area of forest reserves and watersheds in the Conservation District amounted to more than 485,830 ha (1.2 million a) (Hawaii State Department of Planning and Economic Development 1985).

After more than 100 years of tree planting and species trials in Hawai'i, a number of species (particularly *Eucalyptus*) are known that can be successfully grown for timber production (Schubert and Whitesell 1985). However, a significant local timber industry utilizing these nonnative planted trees has failed to materialize, probably because of economic factors, particularly the difficulty that local wood products have in competing with those produced on a vastly larger scale on the U.S. Mainland. Other limiting factors may be the expense of drying and curing wood in high-rainfall areas and the acceptability of locally grown hardwood to those accustomed to using Mainland softwood timber.

Private Commercial Forestry. Commercial forestry projects have been conducted by private landowners as well as by the State. In the past, private lands supporting what was considered to be "commercial forest" exceeded in area such forest owned by the State (Nelson and Wheeler 1963; Hawaii State Department of Land and Natural Resources 1973). A notable example of commercial forest development on private land was that of Hōnaunau Forest of the South Kona District on Hawai'i Island, owned by the Bishop Estate. Between 1956 and 1962, more than 340 ha (850 a) were bulldozed and planted with nonnative hardwoods and pines. Costs for clearing, planting, and maintenance averaged $124/ha ($50/a), and potential earnings were estimated to be $2,960-4,690/ha ($1,200-1,900/a), varying with the value of the different planted species (Carlson and Bryan 1963). The foresters involved with this project optimistically predicted that within 40 years, Hawai'i could be "exporting lumber to the Pacific Northwest." However, such a reversal of lumber importation was not to be. Ten years later, one of the authors of the prior report had revised estimates of clearing costs, investment required, and length of time until harvest. By this time, 810 ha (2,000 a) of Hōnaunau Forest had been planted at a cost of $250/ha ($100/a). This amounted to an investment of $7,660/ha ($3,100/a) over 45 years, approximately double what the Estate could hope to be paid for the harvested timber (Anon. 1973).

Norman Carlson, land manager for Bishop Estate on Hawai'i Island, was quoted as saying that the mistake at Hōnaunau was planting introduced trees rather than the native *koa (Acacia koa)* (Anon. 1973). The Estate attempted to rectify this "mistake" by practicing koa silviculture both at Hōnaunau and on their lands at Keauhou Ranch, upslope of Kīlauea Crater. At Hōnaunau Forest, koa was logged and replanted (1,000 trees/year) between 1927 and 1946 (Carlson and Bryan 1963). Commercial koa silviculture began in the early 1980s, when 60 ha (150 a) of open to closed *'ōhi'a (Metrosideros polymorpha)*/koa forest were bulldozed and planted with koa (Clarke et al. 1980). Reforestation was practiced on a larger scale at Keauhou, where koa/'ōhi'a forest formerly selectively logged and grazed was bulldozed and planted in increments of 20 to 200 ha (50-500 a), beginning in 1976 (Corn et al. 1978; Skolmen and Fujii 1980). The Keauhou site selected for koa silviculture was prime habitat for *Vicia menziesii*, the first endangered plant species listed from Hawai'i, as well as for four endangered bird species: *'io (Buteo solitarius), 'akiapola'au (Hemignathus munroi), 'akepa (Loxops coccineus),* and Hawaii creeper *(Oreomystis mana)* (Warshauer and Jacobi 1982; Scott et al. 1986).

After three years of research on the stocking and growth of koa at Keauhou Ranch, the project was declared a success, because koa growth was rapid, natural reproduction after scarification (supplemented by planting) gave a good stocking rate, and a dense, pure stand of vigorous koa was the result (Skolmen and Fujii 1980). However, a few years later, one of the project's research foresters concluded that large-scale koa silviculture and reforestation would be unlikely in the future because of high clearing costs, the need for long-term investment, and low koa prices. His feasibility study indicated that clearing, fencing, planting, and maintenance of planted koa stands would initially cost $2,500/ha ($1,000/a), which would amount to an investment of $44,500/ha ($18,000/a) over the 50 years required for the stand to mature. Only $15,000/ha ($6,000/a) could be derived from the sale of the koa, resulting in a significant loss for the landowner (Skolmen 1986a).

Koa has always been considered the most valuable of Hawai'i's endemic trees (Nelson and Honda 1966), but logging in natural stands of koa results in a relatively low yield of lumber (1,200-4,900 board feet per hectare or 500-2,000 bd.ft./a), and logging and milling costs are high (Carlson and Bryan, n.d.). Most of the accessible lowland koa forests were logged in the 19th century, and conversion to sugar plantations or cattle ranches has destroyed or degraded much of the koa forest of both logged and unlogged areas (Culliney 1988). Despite the "depleted" status of the koa ecosystem, remaining upland koa forests in private hands continue to be logged, at an estimated removal rate of 10,000 board feet/day on Hawai'i Island (Powell and Warshauer 1985a).

The State holds some koa forest in forest reserves, but this amounted to only 8,910 ha (22,000 a) in 1961, when it was almost all classified as "commercial forest" (Nelson and Wheeler 1963). Currently, most State land supporting koa is leased out to cattle ranchers (Skolmen 1986a). The only modern instance of koa logging without intent to convert to agricultural land use occurred on State land in 1970, when koa timber of the Laupāhoehoe Section of Hilo Forest Reserve was sold under contract to be logged (Skolmen 1986a). Two hundred hectares (500 a) were selectively logged for koa, and some 'ōhi'a was also removed. Though koa seedling germination was stimulated by the logging disturbance, natural replacement of koa in this logged forest was severely hampered by feral pig *(Sus scrofa)* depredations and competition from the alien vine banana poka *(Passiflora mollissima)* (Scowcroft and Nelson 1976). Actions such as contract logging in a forest reserve and later "maintenance logging" of koa on Hawaii Division of State Parks land on Kaua'i (Powell and Warshauer 1985b) only served to further reduce the State's limited koa forest.

While the future of koa silviculture may be clouded in economic uncertainties, interest in growing koa remains relatively high among large landowners (S. Amundson, pers. comm. 1988). It is uncertain whether future development of planted koa stands will be in existing forests (as with the Bishop Estate plantations), or in pasturelands formerly covered by koa forest, a use of less than optimal ranchland proposed by ecologists over 15 years ago (Spatz 1973).

Exploitation of Tree Ferns. Although *koa (Acacia koa)* is the most commercially valuable native plant in the forests of Hawai'i, endemic tree ferns or *hāpu'u* (*Cibotium* spp.) have also been considered an important resource and have been exploited for several decades. The trunks of several species of *Cibotium* produce a coarse fiber that is widely used in Hawai'i as a potting medium and is also exported. About half of the "commercial" forest land on the island of Hawai'i supports tree ferns (Nelson and Wheeler 1963). In the 1950s and 1960s, a small industry developed based on these species (Nelson and Hornibrook 1962); the value of harvested tree fern was $27,000 in 1969 (Nelson 1973), and $135,000 in 1977 (Armstrong 1983). Foresters estimated that more than 40,490 ha (100,000 a) under State (then territorial) control supported harvestable tree ferns, from which the yield of merchantable ferns might approach 1 million cords. However, much of this land was judged to have access too difficult or expensive to exploit the fern (Yamayoshi 1951). In the same report, areas of lands bearing 'ama'u ferns (*Sadleria* spp.), generally smaller species with uses similar to *Cibotium*, were also tabulated, and it was estimated that more than 3 million 'ama'u within easy access grew on government and private lands.

During the 1960s, the State sold hāpu'u harvest rights to small parcels of forest reserve lands prior to clearing and planting with timber species (Fernwood Industries of Hawaii 1971). For decades, private landowners have also harvested or contracted with others to harvest hāpu'u on their lands, generally prior to clearing land for other purposes. Often private lands have been in the State Conservation District when plans for hāpu'u harvest were made; parcels of thousands of acres were sometimes involved (Bryan 1957; Buck 1982). In a recent example, a large land owner proposed to harvest for export to Japan 3 million m^3 (105 million ft^3) of tree fern trunks on private agricultural and conservation-zoned forest of Hawai'i Island. If carried out, this project would have impacted 104 km^2 (40 mi^2) of forested land (Powell 1985).

The effects of tree fern harvesting in intact montane rain forest were studied by Buck (1982). He concluded that although tree ferns re-established after harvest, alien plant invasion was significant and long lasting in areas with uncontrolled harvest techniques. Similar results were obtained in a study of purposeful canopy opening in an *'ōhi'a (Metrosideros polymorpha)*/hāpu'u forest of Hawaii Volcanoes National Park (Burton 1980). When tree fern fronds were artificially removed from study plots, alien plant invasion resulted. Numbers of weed species and their cover increased with the percentage of the fern canopy removed. Although the effects of tree fern harvesting might not be significant if land is to be logged and cleared for some other use anyway, removal of tree ferns from otherwise intact forest can only result in encouragement of alien plant species and eventual forest deterioration.

Woodchipping/Biomass Ventures. In recent years, eucalyptus (*Eucalyptus* spp.) and other alien species originally planted for a timber industry that never fully materialized have been used as wood chips, both for paper production and fuel. Both forest reserve lands and private plantings have been logged to provide wood chips for export and local consumption (Critchlow 1980). Bryan (1977a) described a wood-chipping operation on State-owned lands of Hāmākua on Hawai'i Island, in which chips were shipped to Japan for pulp. The trees were *Eucalyptus*, planted early in the century, and were expected to re-sprout and be ready for reharvest within 10 to 15 years.

In the last decade, there has been increasing interest in developing a biomass industry to produce a source of renewable energy and reduce the State's dependence on imported oil (Hawaii Island Economic Development Board 1988). The focus of modern biomass projects is on plantations of fast-growing introduced trees. While planted stands of eucalyptus and other alien species already occur in the Islands, the existing area is not sufficient to support a large power plant (Yang et al. 1977). However, the potential biomass energy represented by existing plantations is not insignificant. The volume of planted eucalyptus on recently surveyed plantations (1,490 ha or 3,680 a) would, if chipped, equal the energy of more than 300,000 barrels of oil (Buck et al. 1979).

A biomass energy study team from Stanford University and the University of Hawaii concluded that large-scale plantations of biomass-producing eucalyptus should be grown on marginal agricultural land or non-forested grazing land on Hawai'i Island (Yang et al. 1977). They determined that the native koa *(Acacia koa)* and 'ōhi'a *(Metrosideros polymorpha)* were not suitable as biomass crops because of their slow growth and failure to readily resprout. Furthermore, native forests were considered to be unsuitable as sites of biomass plantations because of environmental concerns. Some of the dozen sites evaluated by the study team and rejected because of their native forest cover have subsequently become State Natural Area Reserves (e.g., Kahauale'a, Pu'u Maka'ala, and Pu'u o 'Umi on Hawai'i Island). In 1987, one of the privately owned sites that the team recommended for development became part of the Hakalau Forest National Wildlife Refuge (Mull 1986a) and is acknowledged to be one of the most important habitats for endangered forest birds in Hawai'i (Scott et al. 1986).

A year after the biomass energy study, C. Brewer Co., in collaboration with federal agencies, began to develop a 283-ha (700-a) biomass plantation on lands north of Hilo (Hawaii Island Economic Development Board 1988). About half of the area bulldozed and planted in this experimental project (138 ha or 341 a) was covered by native 'ōhi'a forest in the "Resource Subzone" of the Conservation District (Mull 1982). U.S. Forest Service bioenergy research at this and other trial plantings on the Island determined that, of 28 species tested (including one native, koa), three species of *Eucalyptus* and two leguminous trees were best suited for biomass production. During the study period, the mean growth of *Eucalyptus grandis, E. saligna*, and *E. globulus* was 3 m (10 ft) per year (Schubert and Whitesell 1985).

In recent years native 'ōhi'a forests have been used as sources of wood chips to fuel power-generators as a substitute for sugar cane bagasse, a use that may increase in the future if sugar plantations cease production. Burning of bagasse provided 38% of Hawai'i Island's electricity in 1985 (Powell and Warshauer 1985b). One recent large-scale woodchipping project in the Puna District (on Campbell Estate lands near Kalapana) generated a great deal of controversy and local opposition (Lockwood 1985). The area chipped was considered by some local botanists to be "the best remaining low-land tropical rain forest in Hawai'i and the U.S." (Powell and Warshauer 1985a). In national publications, woodchipping of Hawaiian forests was compared with land-clearing practices of developing countries (Holden 1985). As the author of a recent Sierra Club book on Hawai'i perceived it, "The rarest flora in the United States was being chipped for burning in a power plant" (Culliney 1988). The forest being chipped was more than 1,215 ha (3,000 a) of low-elevation 'ōhi'a rain forest notable for its plant species richness, abundance of native birds, and lack of feral pig damage (Warshauer 1984). After a public outcry over the chipping of native forest, a professor of the University of Hawaii was requested by the landowner and chipping company to provide evidence of the site's "uniqueness." He pointed out that the Kalapana wood-chip site was an intact remnant of a once-extensive lowland vegetation that was "statistically unique" and notable for its geologically recent substrates with all stages of primary rain forest succession in a small area (Mueller-Dombois 1985).

Despite the high value placed on this one lowland forest by botanists and environmentalists, chipping continued on the privately-owned parcel. Ironically, the chipped forest was part of a larger parcel that was subsequently traded to the State and became (in part) Kahauale'a Natural Area Reserve (Mull 1986b). Warshauer (1984) predicted that private forests would continue to be threatened by similar destruction because current tax laws and zoning provide economic incentive to clear forest and convert it to pasture. Powell and Warshauer (1985a) warned that as the sugar industry decreases production, wood chips will become an even more important fuel, and the demand will outstrip the supply of fuel wood from plantations unless more biomass-producing trees are planted on abandoned caneland. This opinion was echoed in a recent study by the U.S. Forest Service, which criticized the State and County of Hawai'i tax policies for not valuing forest cover or encouraging afforestation (Hawaii Island Economic Development Board 1989).

Modern Sandalwood Exploitation. Another recent development in the exploitation of natural forest resources is the revival of the sandalwood trade. In 1988, a Florida investor, who had recently bought a ranch upslope of Kealakekua, South Kona, on Hawai'i Island logged off more than 1,000 old, large-diameter sandalwood *(Santalum paniculatum* var. *pilgeri)* trees and shipped them to China, reaping a profit perhaps as high as $1 million (TenBruggencate 1988a). Logging was completed on the newly purchased ranch despite the concerns and opposition of State Forestry officials and environmentalists (TenBruggencate 1988b). Subsequently, several more owners of large Kona ranches contracted with loggers to exploit their sandalwood resources, represented by trees, perhaps hundreds of years old, scattered in lands used to graze cattle (Ward 1988). Some of the loggers and landowners involved in the sandalwood logging disputed both the amount of profit made and the negative effects of logging on the forest compared to long-term cattle grazing (TenBruggencate 1988b).

At the same time as the modern sandalwood logging was being discussed in Hawai'i newspapers, problems with Hawai'i Island watersheds were the topic of a series of meetings with State legislators and concerned citizens. To many, the revival of sandalwood logging seemed to point out the need for increased protection of existing forests (Anon. 1988). Surprisingly, much of the watershed forest of Puna and Kona Districts is zoned Agriculture and thus could be logged or chipped (Warshauer 1984), despite the land's value as watershed for downslope development.

Ranching

Cattle ranching uses far more land in the Hawaiian Islands than is used for crop production. In 1982, pasture and rangeland in the State amounted to more than 400,000 ha (1 million a), three times the area devoted to cropland, and nearly a quarter million animals were supported on this grazing land (Hawaii State Department of Planning and Economic Development 1985). Land used for cattle ranching constitutes more than 25% of the total land area in the State (Hugh et al. 1986). Even more land was devoted to raising cattle *(Bos taurus)* in the past; in 1960 more than 2 million acres were used for cattle grazing (Schmitt 1977), amounting to 52% of the State's total area (Baker 1961). More than half of some Islands was put to this use. In 1960, 65% of Hawai'i Island and 61% of Moloka'i were used for grazing (Baker 1961). By the 1980s Gagné (1988) estimated that 54% of Hawai'i Island was grazing land. Zoning and tax laws have contributed to the conversion of native vegetation to ranchland in Hawai'i. Much biologically important forest and shrubland was zoned Agricultural before its true value was recognized. If privately owned, even Conservation-zoned forests are not immune to development for ranches or other agricultural uses. Warshauer (1986) commented, "many decision makers consider Conservation District lands as a "garbage-bin" designation, awaiting rezoning and permit granting for other uses." Such removal from the Conservation District and conversion to ranches and macadamia nut orchards was recently carried out at Kapu'a, South Kona, an area known to contain rare plants and the endangered Hawaiian hoary bat *(Lasiurus cinereus semotus)* (Newman 1984) and recognized as a botanically rich and diverse forest early in the century (Rock 1913; Judd 1932). The tax laws in Hawai'i encourage the clearing of forest and conversion to cattle grazing by assessing such "low-grade pasture" at a lower value than uncleared forest, so clearing forest may result in lower taxes and increased real estate values and salability (Warshauer 1984).

Ranching, like large-scale commercial crop production, became firmly established only after land ownership was clarified in the middle of the 19th century. As early as 1831, at least 2,000 cattle, apparently domesticated, were present on the island of O'ahu (Meyen 1981), and by 1851, the number of cattle on O'ahu had increased to 12,000 (Henke 1929). On the island of Hawai'i, the decade of the 1830s marked the beginning of the Parker Ranch (the State's largest ranch), when John Parker began to acquire and lease land and stock it with captured and tamed wild cattle (Wellmon 1969). By the 1850s, the number of cattle (wild and tame) in the Hawaiian Islands was estimated as more than 40,000, and ranchers had begun to import breeding livestock to improve their herds (Henke 1929).

Nearly 40 large ranches were developed in the Islands between 1860 and 1916, more than one-third of them on the island of Hawai'i (Henke 1929). While the ranches of O'ahu, Kaua'i, and West Maui were generally no larger than a few thousand acres and were often associated with lowland sugar plantations, many ranches of Hawai'i and East Maui were tens of thousands of acres in size and frequently extended far upslope. In the mid-20th century, these large ranches produced more than 90% of the State's cattle (Philipp 1953). Much upper-elevation rangeland was essentially native forest when ranching commenced. As late as the 1960s, nearly one-third of the State's grazing land was actually forest, and State forest land was considered by the Division of Forestry to have great potential for livestock production in a multiple-use approach to forest resources (Nelson and Wheeler 1963).

More than half of the land controlled by the State government has been used for grazing in recent times (Culliney 1988). Beginning in the mid-19th century, several large ranches in the drier districts of Hawai'i Island were developed on lands wholly owned by the government. Other ranches, particularly on Maui and Hawai'i, were assembled from private lands and adjacent lands leased from the Territory (and later State) (Henke 1929). While much of this government land was considered "waste" by ranchers (Henke 1929), botanists have long recognized that some of the dry forests contained within such grazing leases are among the most unique and species rich in the Hawaiian Islands (Rock 1913). An example of State-owned ranchland with important dry forests is Pu'uwa'awa'a in North Kona, Hawai'i Island, a parcel of more than 51,820 ha (128,000 a) that supports several unique plant communities and a number of rare or endangered tree and shrub species (Powell and Warshauer 1985b). A preserve in this botanically rich area was suggested early in this century, and finally in 1985 a formal proposal arose to remove 32,390 ha (80,000 a) from grazing and establish a Natural Area Reserve up to several thousand hectares (3,000-12,000 a) in size (Powell and Warshauer 1985a). A fire swept through the proposed reserve in 1986, and, to date, the Pu'uwa'awa'a Natural Area Reserve has not been finalized. Another significant dry forest site on State-owned ranchland of Maui has fared better; this is Kanaio forest, a remnant of formerly extensive dryland vegetation on leeward Haleakalā. Although part of 'Ulupalakua Ranch for more than 100 years, several hundred hectares of native forest and shrubland were deemed valuable enough to be proposed for addition to the State's Natural Area Reserves System (Natural Area Reserves System Commission 1989). Not only rare plants are affected by the leasing of State land for ranching. In South Kona, a State-owned tract, which is habitat for the critically endangered Hawaiian crow or 'alalā (Corvus hawaiiensis), has been leased to the surrounding ranch for use as pasture, and wildlife biologists have been denied access (Anon. 1989a).

State lands have been heavily impacted by cattle even though they were included in the State Forest Reserve System. Although fencing of reserves was an important element in the protection of watersheds in the first few decades of the 1900s (Hosmer 1959; Bryan 1961a), by mid-century many fences separating reserves from ranches and grazing lands had been allowed to deteriorate, and little new fencing was undertaken. The subsequent ingress of cattle and other alien animals has resulted in the degradation of native vegetation and the loss of watershed values in many forest reserves, particularly on the islands of Moloka'i, Maui, and Hawai'i (Warshauer 1986).

The ability of cattle (whether feral or domestic) to degrade vegetation and reduce native forest to grassland pastures was recognized soon after large-scale ranching began during the last century. In 1881, travellers between Waimea and Kukuihaele on Hawai'i Island reported the destruction of a dense koa (Acacia koa) forest by Parker Ranch cattle (Barrera and Kelly 1974). Rosendahl (1976) related accounts of the uplands of Kāne'ohe on windward O'ahu, describing the transformation of hala (Pandanus tectorius) forests into rolling plains used for cattle pastures by the mid-1850s. A similar destruction of hala forests and conversion to a cattle and sheep ranch was reported for the Kahuku region of windward O'ahu (Hosaka 1931). In both windward O'ahu examples, the development of cattle ranching signaled the end of crop cultivation by native Hawaiians. Ranching was also blamed for the demise of the Hawaiian system of field cultivation on *kula* lands in Kona (Kelly 1983) and Waimea (Barrera and Kelly 1974) of Hawai'i Island.

Lowland and mid-elevation forests of dry areas were also affected by cattle at an early date. Frierson (1973) catalogued the historical uses of the Honouliuli *ahupua'a* (land division) of leeward O'ahu, which included the stocking of more than 32,000 cattle on 17,410 ha (43,000 a) of land in 1877. These cattle (along with numerous feral goats, *Capra hircus*) were instrumental in the destruction of almost all trees between 210 and 550 m (700-1,800 ft) elevation. The dry forests of Lāna'i have been reduced to small remnants of their former range through the actions of domestic cattle and sheep (*Ovis aries*) as well as feral goats (Skottsberg 1953; Spence and Montgomery 1976). Partial fencing may have prevented the destruction by cattle, goats, and deer (*Axis axis*) of Kānepu'u, Lāna'i, a botanically rich *lama/olopua (Diospyros sandwicensis/Nestegis sandwicensis)* forest harboring several endangered and otherwise rare dry forest plants (Ziegler 1989). Some mid-elevation forests of leeward Maui (Kula) have been completely destroyed by cattle; others have been much reduced, with forest edges receding upslope (Medeiros et al. 1986). Dry leeward cattle pastures of the saddle region of Hawai'i Island are a further example of the nearly complete conversion of native woody vegetation to alien grasslands. Only the remnant native shrublands of scattered steep-sided cinder cones with their occasional *māmane (Sophora chrysophylla)*, sandalwood *(Santalum ellipticum)*, *kōlea (Myrsine lanaiensis)*, and 'akoko *(Chamaesyce olowaluana)* trees indicate the nature of the original plant cover of this ranchland (Cuddihy et al. 1982a).

The role of cattle in opening up native forests and reducing their value as watersheds was recognized and decried by a number of foresters in the early decades of the 20th century (Koebele 1900; Giffard 1918; Judd 1918). Judd (1927) estimated that only 25% of the Hawaiian Islands was still covered by forest in the late 1920s, a small remnant of the original distribution before cattle. In these early reports of damage, the destructive actions of cattle most often discussed were the trampling of the forest understory (Hall 1904) and the spread of alien grasses such as Hilo grass *(Paspalum conjugatum)* (Giffard 1918). The loss of understory and conversion to pasture sometimes happened very quickly. Campbell (1920) reported the disappearance of tree ferns and other understory plants in a koa forest near Volcano, Hawai'i Island, after less than 25 years of grazing. The loss of understory has been blamed for the drying of the soil and the death of shallow-rooted tree species (Hall 1904; Rock 1913). Long-term cattle grazing in upland forests may result in the complete loss of all but the overstory tree species and the conversion of closed forest into artificial parklands of scattered native trees in alien grass pastures; this situation is graphically illustrated on the upper portions of the recently-established Hakalau Forest National Wildlife Refuge on the slopes of Mauna Kea, an area long used as a cattle ranch before its acquisition as a preserve (Stone et al., in press).

In addition to numerous observations of negative impacts of cattle on natural vegetation in Hawai'i, quantitative studies have documented the changes in vegetation caused by grazing. Baldwin and Fagerlund (1943) analyzed the reproduction of koa in grazed montane parkland vegetation and exclosures protected from cattle. They found almost complete suppression of both seedlings and suckers in the grazed area. In a later study of the same area, cattle were shown to have altered the floristic composition, encouraged the establishment of alien grasses, and reduced the shrub cover and height in grasslands and shrub-dominated communities (Cuddihy 1984). Once alien grasses become established in native forest or woodland used for cattle grazing, they may comprise significant cover for many years following removal of animals. An example is the koa parkland upslope of Kīpuka Puaulu in Hawaii Volcanoes National Park, where alien grasses such as *Paspalum* spp. and meadow ricegrass *(Ehrharta stipoides)* are still prominent 40 years after cessation of grazing. Although cattle were fenced out of portions of the diverse mesic forest of Kīpuka Puaulu as early as 1926 (Apple 1954), heavy grazing resulted in the loss of almost half the native tree species between 1913 and 1967 (Mueller-Dombois and Lamoureux 1967).

Cattle are known to significantly reduce tree cover in montane māmane forests. Scowcroft (1983) recorded a loss of one-fifth of the tree cover in cattle-grazed māmane forests of Mauna Kea over a period of 21 years, probably due to suppression of reproduction as well as direct loss of cover on mature trees. On East Maui, cattle browse preferentially on certain native woody species such as *kuluʻi (Nototrichium sandwicense)* and also break branches and trample roots of less common endemics such as *hōlei (Ochrosia haleakalae)* and sandalwood *(Santalum freycinetianum* var. *auwahiense)* (Medeiros et al. 1986).

Given enough time and a high enough stocking rate, cattle are capable of destroying native vegetation once they are introduced to an area. However, associated human activities have generally intensified and accelerated destruction. In the past, ranch timber was typically cut for use in fencing or construction of ranch buildings. Often logging (particularly of koa) preceded ranching operations (Jenkins 1983) or was used to expand existing pastures (Warshauer and Jacobi 1982). Since native shrubs such as *pūkiawe (Styphelia tameiameiae)* and *ʻaʻaliʻi (Dodonaea viscosa)* are considered to be "noxious plants" on Hawaiian ranges (Hosaka and Thistle 1954), they have often been mechanically cleared to encourage pasture grasses and facilitate stock movement (Apple 1954).

Numerous alien grasses and legumes have been introduced to provide forage for cattle (Hosaka and Ripperton 1944; Whitney et al. 1964), and some of these, such as kikuyu grass *(Pennisetum clandestinum)*, have proven to be invaders of natural vegetation near pastures (Smith 1985). After kikuyu grass invaded an exclosure at Auwahi, East Maui, built to protect rare dry forest tree species, no natural tree reproduction was observed (Medeiros et al. 1986). Some other grasses that have become serious problems in native ecosystems were probably introduced accidentally in animal feed or imported seed; an example is broomsedge *(Andropogon virginicus)*, a pest widespread on the islands of Oʻahu and Hawaiʻi (Sorenson 1977).

Although beef cattle are the focus of the modern livestock industry in Hawaiʻi, sheep ranching was formerly important, at least on some of the Islands. Commercial sheep production began in the Islands in the 1840s and 1850s and peaked in 1884 with 121,000 animals (Henke 1929). Principal sites of sheep ranching were the islands of Lānaʻi and Niʻihau and the Humuʻula Sheep Station of Mauna Kea on Hawaiʻi Island. The impact of sheep on the low and dry island of Niʻihau must have been tremendous. When St. John made botanical collections there in the mid-20th century, only a few native plants were located, remnants of plant communities almost completely destroyed (St. John 1959). The Humuʻula Sheep Station reportedly had 9,000 sheep in 1873 (Bird 1966) and 12,000 head in 1929 (Henke 1929). This herd was the primary source of the feral sheep population of Mauna Kea, which had built up to at least 40,000 animals by 1937 (Tomich 1986).

The island of Kahoʻolawe was used to raise sheep periodically between 1859 and 1918 and was stocked with as many as 20,000 animals in the 1870s. The impact of these domestic sheep, along with cattle and feral goats, was devastating to the dryland vegetation, and the result was the almost total loss of plant cover and subsequent loss of soil by wind erosion, resulting in a hardpan substrate on the upper third of the Island (Myhre 1970). The vegetation and soils of Lānaʻi were also exposed to domestic sheep, with a herd of up to 50,000 animals at times in the early 1900s (Henke 1929). The result, as on Kahoʻolawe, was widespread loss of native woody vegetation and soil erosion, which George Munro, a Lānaʻi land manager and authority on Hawaiian birds, attempted to ameliorate by tree planting in the first few decades of the 20th century (Spence and Montgomery 1976). By the 1960s, only a few sheep were still being raised on the islands of Hawaiʻi and Niʻihau (Tomich 1986), and by the mid-1980s sheep numbers were too insignificant to be reported in the State Data Book (Hawaii State Department of Planning and Economic Development 1985).

Feral Ungulates

Feral ungulates (domesticated hoofed mammals gone wild) have been extremely important causes of vegetation decline in Hawaiʻi. The only ungulate introduced by Polynesian peoples, the Polynesian pig *(Sus scrofa)*, probably produced only minor damage to native vegetation because of its close association with settlements in the lowlands. There is little evidence that upland populations of Polynesian pigs penetrated native forests.

Of the early European domesticated introductions that became feral, cattle, goats, pigs, and sheep *(Bos taurus, Capra hircus, Sus scrofa, Ovis aries)* were most destructive to native ecosystems. Other more recently introduced ungulates such as mouflon sheep, axis deer, and mule deer *(Ovis musimon, Axis axis, Odocoileus hemionus)* have also degraded forests in certain areas. Effects of the ranching industry and the negative impacts of feral cattle on native vegetation have already been discussed herein.

Foraging and trampling by goats, pigs, and other ungulates can result in severe erosion of watersheds, because these mammals inhabit terrain that is often steep and remote. As early as 1900, there was increasing concern about the integrity of island watersheds, and a professional forestry program emphasizing soil and water conservation was begun (Holt 1989). Feral ungulates were removed from forested areas, forest reserves were established, and reforestation was begun. A recent attempt to re-emphasize watershed protection in the Hawaiian Islands was the introduction of a bill to the Hawaiʻi State Legislature in 1989, including a measure to reduce ungulates on important watersheds. Unfortunately the bill died in committee after opposition from sport hunters.

Street (1989) gave examples of ungulate-caused erosion in Hawaiʻi. Exposed pink rock in the windward valleys of Oʻahu; splotches of red in Waimea Canyon on Kauaʻi; deep, red gullies and silting fishponds on Molokaʻi; and exposed, hardened subsoils of Kahoʻolawe are ungulate effects. Concern about the degradation of watersheds by pigs, goats, and deer has been especially emphasized on Maui in discussions about water quality and disease (Hills 1988), and on Molokaʻi in connection with sedimentation of coastal fishponds. Weisler and Kirch (1982) estimated that erosion from the uplands of Kawela, Molokaʻi, has deposited about 0.3 m (1 ft) of sediment per year over the last 100 years in fishponds. They blamed feral ungulates for the erosion. Goats and sheep have been responsible for considerable wind erosion of Kahoʻolawe, and although ungulate removal was recommended as early as 1910 (Myhre 1970), reclamation has not yet been successful.

Feral Goats. Goats *(Capra hircus)* were first introduced to Niʻihau by Capt. Cook on February 2, 1778, and may have been set ashore on the island of Hawaiʻi soon after (Tomich 1986). Captain Vancouver introduced additional goats to Hawaiʻi Island starting in 1792. By the 1820s, goats had reached remote areas on Oʻahu and Kauaʻi and probably several other Islands. On Oʻahu alone, where records were most complete, an average of 50,000 goat skins per year was shipped out of Honolulu between the 1840s and 1880s, and goat husbandry was thriving during that period (Culliney 1988). The peak year for goat skin production was 1855, when more than 100,000 were exported from the Islands (Schmitt 1977). The goat population on the island of Hawaiʻi may have been about 75,000 in 1931 (Bryan 1931).

Able to reach more remote and inaccessible areas than other ungulates, thriving on a variety of food plants, and often continuing damage begun by cattle, goats were and are important in the decline of native vegetation in many areas. They now occur on all main Islands except Niʻihau and Lānaʻi, where they occurred in the past. By 1920, goats were greatly reduced on Oʻahu, where they are scarce today (Tomich 1986).

On Maui and Hawaiʻi, feral goats damage subalpine woodlands and alpine grasslands. Medeiros et al. (1986) reported that goat impacts on native plants in the *koa/ʻōhiʻa (Acacia koa/Metrosideros polymorpha)* zone of the south slope of Haleakalā are leading to the loss of forest, resulting in continuing deterioration of watersheds. In the past, goats destroyed wet forest remnants on Lānaʻi, and they still enter upper-elevation wet forest on Maui. On Molokaʻi, Hawaiʻi, and Maui, goats degrade low-elevation dry forest, and on Kauaʻi they enter wet habitats such as the Alakaʻi Swamp in dry periods (Stone 1985). Goats once played a major role in destruction of dry and mesic forest on Niʻihau and Kahoʻolawe (Scott et al. 1986; Tomich 1986).

Feral goats are known to limit *māmane (Sophora chrysophylla)* reproduction in the subalpine zone of Haleakala National Park and Mauna Kea (Loope and Scowcroft 1985), and koa reproduction in the montane parkland of Hawaii Volcanoes National Park on Hawaiʻi Island (Spatz and Mueller-Dombois 1973). Goats are probably responsible for reducing populations of several native woody species and grasses in the lowlands of Hawaii Volcanoes National Park (Mueller-Dombois and Spatz 1975). The distribution and composition of such communities can be changed through long-term foraging of goats. Rare plants such as the Kaʻū and Mauna Kea silverswords *(Argyroxiphium kauense* and *A. sandwicense* subsp. *sandwicense)*, the Hawaiian jackbean *(Canavalia kauensis)*, and a number of Haleakalā species (e.g., *Stenogyne microphylla, Schidea haleakalensis, Plantago princeps)*, are threatened by feral goats and may be eliminated from heavily used areas. Fortunately, systematic hunting and drives, combined with barrier fencing and the "Judas goat" technique using radio-collared animals (Taylor and Katahira 1988), have been successfully used to reduce and eliminate feral goats in many important natural areas.

Feral Pigs. The European domestic pig *(Sus scrofa)*, like the domestic goat *(Capra hircus)*, was introduced to Niʻihau during Captain Cook's first voyage, February 2, 1778. Wild pigs were present on Kahoʻolawe before 1840, but these may have been partly of Asian ancestry (Tomich 1986). They disappeared from that island at a later date. Feral pigs were introduced to Lānaʻi in 1911 when a piggery was established in the Pālāwai Basin. They were removed to protect the watershed by the 1930s (Ziegler 1989). Pigs occur on Niʻihau, Kauaʻi, Oʻahu, Molokaʻi, Maui, and Hawaiʻi. Their abundance and distribution in the uplands may have begun with the ten-year *kapu* (taboo) in the late 18th century, but in at least some areas (e.g., East Maui and some parts of West Maui) range expansion or intensification of numbers continues in the mid to late 1980s.

On Hawaiʻi Island, feral pigs are found from dry coastal grasslands through rain forest and in the subalpine zone of Mauna Kea and Mauna Loa. On Maui, Kauaʻi, Oʻahu, and Molokaʻi they inhabit rain forests and grasslands. Although pigs may still be seen in the Waiʻanae Mountains and some residential areas are being invaded, numbers on Oʻahu are no longer large, as a result of reduced habitat and increased human populations (Tomich 1986). Densities in rain forests are generally higher than in all other vegetation types, probably as a result of abundant water and food there.

Feral pig damage to native vegetation has reached extreme levels in the 20th century, probably because of an increase in pig densities and expansion of their distributions. Additional available animal protein in the form of alien earthworms (and possibly other invertebrates), certain recently available invasive alien plants such as strawberry guava *(Psidium cattleianum)* and banana poka *(Passiflora mollissima)*, increasingly abundant disturbed areas created by humans and livestock, favorable climatic trends, and even reduction of predation by hunters and feral dogs *(Canis familiaris)*, may have accelerated population growth and expansion into additional areas (Stone 1985).

Large areas of native forest can be affected by pig activity. It has been estimated that about one-third of the diggable area in a mountain rain forest on Hawaiʻi Island can be disturbed in a year's time by these animals (Cooray and Mueller-Dombois 1981). Large expanses of such disturbed soil provide fertile seedbeds for alien plants, some of them spread through ingestion and excretion by pigs. In one study in Hawaii Volcanoes National Park, seeds of the alien firetree *(Myrica faya)*, a serious pest, constituted an average of 5% of the stomach contents of pigs removed from a fenced rain forest (Stone and Taylor 1984). Increased nutrient availability resulting from pig activity in nitrogen-poor soils further facilitates establishment of alien weeds more adapted to richer soils than native plants. Open and largely alien forests can completely replace native stands.

Pigs also destroy seeds and seedlings of such dominant native species as *koa, māmane,* and *pilo (Acacia koa, Sophora chrysophylla, Coprosma* spp.*)*, as well as tree ferns or *hāpuʻu (Cibotium* spp.*)* and *ʻamaʻu (Sadleria* spp.*)*. They trample numbers of rare native plants such as *Cyrtandra, Stenogyne,* and *Phyllostegia* (Diong 1982; Stone 1985; Stone, in press). Bogs and the rare plants in them seem especially vulnerable to pig depredations (Loope et al., in press a).

Feral Sheep. Sheep *(Ovis aries)* were apparently first introduced to Hawaiʻi by Captain James Colnet in April 1791 on Kauaʻi (Tomich 1986). Captain Vancouver introduced sheep to Kawaihae on Hawaiʻi Island in February 1793 and shortly thereafter to Kealakekua to the south. More animals were given to Kamehameha I at Kealakekua in January of 1794, and some additional animals may have been left on other Islands that year. Like cattle *(Bos taurus)*, sheep from these early releases were protected by *kapu* (taboo) (Tomich 1986). However, sheep apparently did not do well in the wild until large-scale ranching began in the late 1800s; possibly they were held in check by

65

wet habitats and feral dogs *(Canis familiaris)*. Ellis (1969) speculated about the impact of wild dogs on sheep on Mauna Kea about 1825. The altitudinal distribution of sheep on Hawai'i formerly covered dry coastal areas at sea level through dry mesic shrubland and forest to 3,960 m (13,000 ft) in the alpine zone of Mauna Kea.

Lāna'i, Hawai'i Island, and Ni'ihau were important early sites for sheep production. On Lāna'i, sheep ranching lasted from the mid 1800s until about 1920 (Culliney 1988). On Hawai'i, the Humu'ula Sheep Station was established on leeward Mauna Kea in 1876. The feral sheep population on Mauna Kea reached about 40,000 animals by the mid 1930s, and damage to the forest was severe, especially near tree line (Bryan 1937; Warner 1960). Feral sheep were concentrated in *māmane (Sophora chrysophylla)* forests on Mauna Kea, where they not only browsed the leaves and shoots of māmane, but also stripped off bark, which killed or weakened established trees (Scowcroft and Sakai 1983). In addition to direct effects on māmane, sheep have altered the composition and distribution of Mauna Kea forests. *Naio (Myoporum sandwicense)*, being less palatable to sheep than māmane, displaced māmane in heavily browsed areas. Other woody species such as *'akoko (Chamaesyce spp.)* and *na'ena'e (Dubautia)* have become restricted to rocky and steep areas because of browsing of sheep and cattle. Sheep on Mauna Kea were not seriously controlled until 1934 (Tomich 1986), when territorial foresters were able to reduce the sheep population. With public hunting begun in 1955, the numbers were kept below 5,000 animals, averaging 1,500 during the 1970s (Scowcroft and Giffin 1983). Still, consumption of the dominant forest species upon which the survival of an endangered honeycreeper (Drepanidinae), the *palila (Loxioides baileui)*, depends resulted in Federal court orders in 1979 and 1981 to completely remove feral sheep from Mauna Kea. Animals were considered eradicated in July 1981, although repopulation below a "deteriorated boundary fence" is a threat (Tomich 1986).

On Ni'ihau, 15,000-20,000 head of sheep were present by the early 1870s, and up to 30,000 animals still exist (Culliney 1988). Sheep on Kaho'olawe and Ni'ihau were considered partly responsible for destruction of much of the native vegetation (Wagner et al. 1985). Feral sheep have not been seen on Kaho'olawe since the early 1980s (R.L. Walker, pers. comm. 1984).

Mouflon Sheep. Mouflon *(Ovis musimon)*, established on Lāna'i and Hawai'i Island, have effects on native vegetation similar to those of feral sheep *(O. aries)* (Giffin 1982), although usually they travel in smaller groups. Mouflon were released on Lāna'i in July 1954, presently inhabit dry *kiawe (Prosopis pallida)* forests on the western end of the Island, and apparently do not penetrate native forest (Scott et al. 1986). Mouflon were first released on Mauna Kea beginning in 1962 and inhabit *māmane (Sophora chrysophylla)* forests there. They were also released in the Ka'ū District of Hawai'i Island, where they inhabit montane and subalpine *'ōhi'a (Metrosideros polymorpha)* vegetation at 1,220 to 2,800 m (4,000-9,180 ft) elevation and alpine scrub vegetation. On Mauna Kea mouflon threaten the endangered Mauna Kea silversword *(Argyroxiphium sandwicense* subsp. *sandwicense)*, and in the Ka'ū District they threaten the Ka'ū silversword *(A. kauense)*. The State of Hawai'i was ordered by Federal courts to remove mouflon from Mauna Kea in 1986 and 1988.

Axis Deer. Axis deer *(Axis axis)* were first released on Moloka'i in 1868. They were also introduced to O'ahu sometime in the late 1800s and to Lāna'i in 1920. Perkins (1903) commented that between 1896 and 1902, a forest on Moloka'i lost "two-thirds or more of all its trees to cattle and axis deer." The deer on Moloka'i were certainly pests within 30 years of introduction, but since 1956 a large portion of the herd has been managed by the State. The Oʻahu deer persist in small numbers in *kiawe (Prosopis pallida)* thickets above Salt Lake; the Lānaʻi herd of a few thousand has been managed by the State since 1959. By 1960, axis deer had been released on Maui, and the population seems to be extending its range to some degree in dry alien forest (Tomich 1986). The State Division of Game proposed releasing axis deer on Hawaiʻi Island in the 1960s and again in the 1970s, but opposition from farmers and ranchers and problems with environmental assessment requirements prevented the introduction (Juvik and Juvik 1984).

Axis deer are viewed as important in degrading the vegetation and soils of East Molokaʻi, with attendant siltation of coral reefs (Scott et al. 1986). They are also seen by some as a serious threat to Lānaʻi's remaining forests and a threat to mesic and rain forests on Haleakalā on Maui (although not yet present there). Creation of open forest by cattle, pigs, and goats *(Bos taurus, Sus scrofa, Capra hircus)* could facilitate invasion by axis deer (Stone 1985).

Mule Deer. Mule or black-tailed deer *(Odocoileus hemionus columbianus)* from Oregon were first introduced to western Kauaʻi public lands in June 1961. Deer now inhabit mesic and dryland alien forest at elevations of 0 to 1,220 m (0-4,000 ft) (Telfer 1988). The population of perhaps 350 animals is probably kept in check by illegal kills (Telfer 1982, 1988). *Koa, pilo,* and *ʻōhiʻa (Acacia koa, Coprosma* spp.*, Metrosideros polymorpha)* are among the native species eaten in small quantities by deer in largely alien plant habitat. Some of the best rare bird and plant habitat remaining in the State lies in the Alakaʻi Swamp of Kauaʻi, which is invaded to some extent by deer during dry cycles. It is recommended that deer be managed for minimum populations in native forest areas (Telfer 1988).

Rodents

Introduced rats and mice can affect native vegetation through interruption of plant life cycles. Direct damage to plant propagules, seedlings, or native trees by rodents, if frequent enough, can change forest composition and structure (Campbell 1978). Indirect effects of rodents through consumption of significant numbers of plant pollinators (insects or birds) can also eventually influence the distribution and abundance of plant species. Rodents may also incidentally pollinate native plants (e.g., *ʻieʻie* or *Freycinetia* and *loulu* palm or *Pritchardia*) or alien plants in Hawaiʻi. The effects of rodents during irruptions when populations are high can be considerably greater than at other times (Atkinson 1977). Initial establishment of rodents in an area is often followed by higher numbers of animals than can be sustained over the long term.

Four species of rodents have established populations in Hawaiʻi. Only the black or roof rat *(Rattus rattus)* has been recorded from Niʻihau; the Norway rat *(Rattus norvegicus)* has been recorded from all main Islands except Kahoʻolawe. All four species of rodents, including the Polynesian rat *(R. exulans)* and the house mouse *(Mus domesticus)*, are present on Kauaʻi, Oʻahu, Molokaʻi, Lānaʻi, Maui, and Hawaiʻi Islands.

House Mouse. Mice *(Mus domesticus)* were introduced to Hawaiʻi by 1816 (Kotzebue 1821) and now range from sea level to about 3,960 m (13,000 ft) elevation (van Riper and van Riper 1982; Tomich 1986). They are abundant over a wide range of vegetation types. Mice are perhaps most common in lowland habitats and are known to sometimes irrupt in "drier beach, grassland, scrub, and forest areas, especially on Maui and Hawai'i" (Tomich 1986; R.L. Walker, pers. comm.). Tomich (1986) reported dense populations in lower elevation wet forests. On Maui, substantial populations were found in some wet forest areas above 1,190 m (3,900 ft) in elevation but not in others (C.P. Stone, unpubl. data). Little is known about mouse food habits in Hawai'i, except that in the lowlands, insects, grass seeds, and fruits are taken (Kami 1966). Baker (1979) stated that mice ate large seeds of the endangered *hau-kuahiwi* (*Hibiscadelphus giffardianus*) on the ground in mesic forest at about 1,200 m (4,000 ft) elevation. Cole et al. (1986a) found that mice in high-elevation shrublands in Haleakala National Park were opportunistic feeders. Diets consisted mostly of grass seeds and various fruits when abundant but shifted to arthropods when these foods were scarce. Cole et al. (1986a) noted that disruption of plant pollination through destruction of larvae of pollinators by mice is a distinct possibility. Mouse populations at high elevations appear to be more subject to fluctuation in response to seasons and weather patterns than rats; at peak population levels mice may be very destructive of arthropods.

Rats. Polynesian rats (*Rattus exulans*), because they were introduced so long ago, likely influenced native vegetation composition in some areas of Hawai'i. Their possible past impacts were discussed with those of other Polynesian introductions, and it was noted that this species may now be more abundant than previously believed at elevations up to 1,980 m (6,500 ft) elevation. Studies of the ecology of Polynesian rats are being conducted at present (C.P. Stone, unpubl. data).

Norway rats (*Rattus norvegicus*) are found primarily around human habitations, and their role in the alteration of Hawaiian vegetation is probably minor. In contrast, black rats (*Rattus rattus*) are widely distributed ecologically, although they are especially common at low to mid elevations (Tomich 1986). Black rats are arboreal and able to remove fruits from small branches and girdle tender tree limbs and trunks, as well as forage for plant propagules on the ground. They are predators on native birds, some of which may once have pollinated rare or extinct Hawaiian plants. Rats are thought to sometimes pollinate 'ie'ie (*Freycinetia arborea*) vines (Perkins 1903; Stone 1985). However, Perkins (1903) noted the difficulty of finding 'ie'ie fruits uncontaminated by rats on O'ahu. Black rats were probably established in the Islands between 1870 and 1890 (Atkinson 1977, 1985).

Black rats strip bark from *koa* (*Acacia koa*) saplings, thus inhibiting or stopping regrowth (Scowcroft and Sakai 1984). Korte (1963) and Whitesell (1964) reported rat problems in reforestation with koa. Rat damage has been observed on stems of the endangered *Vicia menziesii*, a vine in the pea family (Clarke et al. 1983). Black rats damage flowers, fruits, and bark of the rare *Hibiscadelphus* (Baker and Allen 1976; Baker 1979; Russell 1980) and bark of a'e (*Zanthoxylum dipetalum*), olopua (*Nestegis sandwicensis*), pilo (*Coprosma rhynchocarpa*), and hō'awa (*Pittosporum* spp.). Saplings or small trees may be girdled and killed.

Rats probably favor woody plants with large seeds. J. Jacobi (pers. comm.) reported that 75% of the immature fruits of hō'awa were opened by rats in three instances in a Hāna rain forest study; he speculated that the hō'awa population could be drastically reduced over time. Rats regularly eat all *Pittosporum* seeds in a 40-ha (100-a) *kīpuka* (island of older vegetation) in Hawaii Volcanoes National Park and therefore prevent reproduction of the species there. Baker (1979) estimated that 80-90% of the large seeds on individual *Hibiscadelphus* trees could be destroyed in mesic forest at 1,190 m (3,900 ft) elevation. Damage to ʻieʻie, hōʻawa *(Pittosporum hosmeri)*, sandalwood *(Santalum paniculatum)*, and a number of alien trees has been noted (Perkins 1903; Kami 1966; Atkinson 1977). Medeiros et al. (1986) reported rat damage to *māhoe (Alectryon)*, ʻalaʻa *(Planchonella)*, and halapepe *(Pleomele)* on the south slope of Haleakalā. Heavy depredations of black rats on introduced coconuts *(Cocos nucifera)* in many countries is common (e.g., Twibell 1973), and rat damage to the endemic palm *loulu (Pritchardia* spp.*)* was once also severe in some areas of Hawaiʻi (Bryan 1915).

In rain forests, black rats eat plant material more frequently than they eat adult or immature insects (based on limited sampling). In mesic forests, black rats consume green plant materials more frequently than insect larvae and seldom seem to eat adult insects. Seeds dominated in summer diets and green plants in winter in one study (Stone 1985). A.C. Medeiros (pers. comm.) noted that the few fruits of the Haleakalā sandalwood *(Santalum haleakalae)* that reach maturity in the subalpine zone of Haleakalā are subject to heavy rodent predation (by black rats and mice).

Campbell (1978) suggested that trees that evolve in the presence of seed predators may produce all their seeds over a short time period with irregular intensity or extend their fruiting periods. Many Hawaiian plants presumably evolved in response to bird predation since no small mammal seed predators were present until introduced in the 4th century by Polynesians. Unfortunately, extended fruiting periods may now sustain high numbers of rats.

In New Zealand, rodents prey on larger invertebrates, and the absence and disjunct distributions of many invertebrates is attributed directly to predation (Ramsay 1978). Beetles, moths, centipedes, slugs, and snails are among the groups affected. Flightless forms of insects are particularly vulnerable and are now most abundant on offshore New Zealand islands, where rats are absent. In Hawaiʻi, flightless crickets (Orthoptera) and earwigs (Dermaptera) on Nihoa, which is presumably still without rats, present a parallel example (Conant et al. 1984). Ramsay (1978) noted that Polynesian rats feed on relatively large invertebrate prey in New Zealand, those at least approximately 20 mm (0.8 in.) long as adults.

In Hawaiʻi, arthropod larvae occurred in over 80% of all black rat stomachs sampled from mesic forests, and adult insects occurred in 30%. In rain forests, insect larvae occurred in 17% of the stomachs sampled, and adult insects in 32%. Small sample sizes and difficulty of identification preclude statements about pollinators taken by rats (Cole et al. 1986a; C.P. Stone, unpubl. data). Polynesian rats apparently take less vegetable material than black rats in Hawaiian rain forests (F.G. Howarth, pers. comm.). However, their effects on invertebrates such as spiders and crickets in the leaf duff and litter is especially important. Some forms such as earthworms that are involved in nutrient cycling may be affected.

Campbell (1978), through use of rat exclosures in New Zealand, showed that rats can affect species composition of forest plants over several years. Polynesian rats affected the composition of regenerating coastal forest, and black rats were major predators on podocarp seeds in mainland forests. Campbell noted that seeds carried away from fruiting trees are more important in determining forest composition than seeds that fall under trees; the latter often do not survive. Predation of rats on birds that formerly dispersed seeds can also be an influence on forest ecosystems. Effects of Polynesian and black rats on forest composition in Hawaiian rain forests are unknown at present, but the great abundance of these small mammals suggests that vegetation changes have resulted from their activities.

Alien Birds

Introduced birds can be an important means of distributing introduced and native plants, sometimes over long distances. Since 1850, more than 52 species of nongame birds (species not hunted or classified as game birds) and 78 species of game birds have been released in the Islands, of which probably 15 species of game and 30 nongame birds remain. Game birds are known to disperse native and alien plants (Burr 1984). For example, alien ring-necked pheasants (*Phasianus calchinus*) and chukars (*Alectaris chukar*) ingested fruits of available native shrubs (*Vaccinium reticulatum, Styphelia tameiameiae, Coprosma* spp., and *Geranium cuneatum*) in high-elevation shrubland in Haleakala National Park and thereby enhanced germination and probably distribution (Cole et al. 1986b). Lewin and Lewin (1984) noted that the alien banana poka (*Passiflora mollissima*) is the main food of the alien kalij pheasant (*Lophura leucomelana*) on the island of Hawai'i, although many other native and alien plants are also eaten. They noted that many seeds appeared unharmed in the large intestine and feces of this bird. Lewin and Lewin (1984) stated that kalij pheasants "apparently have the ability to enhance the establishment of exotic plant pests."

Alien nongame birds can also spread alien and native plants over considerable distances. The rapid spread of the alien shrub *Lantana camara* through lowland Hawai'i has been attributed to the abundant alien spotted dove (*Streptopelia chinensis*) and common myna (*Acridotheres tristis*) (Perkins and Swezy 1924; Fisher 1948). Similarly, Japanese white-eyes (*Zosterops japonica*), house finches (*Carpodaceus mexicanus*), and other species have likely increased dispersal and intensification of the invasive alien firetree (*Myrica faya*), often found under 'ōhi'a trees (*Metrosideros polymorpha*) in which these birds have perched. Van Riper (1980) speculated that the expansion of the native *naio (Myoporum sandwicense)* on Mauna Kea may be partly the result of alien species such as the red-billed leiothrix (*Leiothrix lutea*), house finch, and turkey (*Meleagris gallopavo*).

Introduced birds are known to irrupt to very high population levels, especially during establishment. For example, populations of the red-billed leiothrix and the common myna were once much higher, and these species were found in different areas in Hawai'i (lowlands and dense forests) than at present (Scott et al. 1986). During peak populations, irruptive alien birds may especially affect the arthropod pollinators of native Hawaiian plants.

Management of introduced game birds includes techniques for "habitat improvement," such as planting alien species of herbs and grasses for bird food and cover and opening of forest and shrublands by bulldozing (Culliney 1988). If native vegetation is subjected to such management, invasion of alien plants and animals will be encouraged and the integrity of native plant communities compromised. Habitat improvement of State and private lands for game birds still occurs (Hawaii State Department of Land and Natural Resources 1980, 1987), but on State lands at least, the scale is reduced and areas of disturbance are confined to alien vegetation (T. Lum, pers. comm. 1989).

Alien Invertebrates

Introduced invertebrates can affect the composition and structure of vegetation directly and indirectly. Invertebrates can feed directly on plants or plant propagules and prey on pollinators of plant species. They also carry diseases that infect native plants and invertebrates; they may support higher levels of alien predators, which in turn deplete native invertebrates more effectively than would otherwise be possible; and they may hybridize with native forms (Howarth 1985).

Alien insects feed directly on endemic plants, sometimes doing extensive damage (Howarth 1985). Defoliation of *māmane (Sophora chrysophylla)* by the moth *Uresiphita polygonalis* and the depredations of the plant louse *Psylla uncatoides* on *koa (Acacia koa)* can seriously disturb the stands involved. Many introduced invertebrates combine in lowland habitats to inhibit koa regeneration. Generalists such as aphids, whiteflies, scale insects, and termites may reduce the geographic ranges of some plant species, and rare plants may be especially jeopardized (Howarth 1985). The black twig borer *(Xylosandrus compactus)* and the associated fungus *Fusarium solani* affect 108 plant species in Hawai'i, some of them rare (Howarth 1985). Sohmer (1976) reported that the black twig borer completely destroyed all *pāpala (Charpentiera obovata* and *C. tomentosa)* shrubs and trees in Kalua'a Gulch, O'ahu, within a period of just six years. These two plants had formerly been abundant in the gulch. The introduced Australian fern weevil *Syagrius fulvitarsus* killed the endemic fern *Sadleria cyatheoides* over large areas on O'ahu and Hawai'i Island until introduction of a braconid parasite *(Doryctes syagrii)* in the 1920s (Pemberton 1964).

Perhaps a thousand species of Hawaiian insects visit flowers and are potential pollinators, including yellow-faced bees *(Hylaeaus* spp.), moths, flies, beetles, and wasps. Native tephritid flies, possibly threatened by a planned eradication program for alien tephritids (e.g., medflies, *Ceratitis capitata*), need investigation (L.L. Loope, pers. comm.). The European honey bee, *Apis mellifera*, with a social foraging strategy unknown in native Hawaiian pollinators, may have reduced outcrossing of some native plants and increased outcrossing in others. Introduced carpenter bees *(Xylocopa sonorina)* and big-headed ants *(Pheidole megacephala)* are known to rob nectar from plants, reducing the chances of pollination and visits by native pollinators (Howarth 1985).

Perkins (1913) noted the devastating effect of the big-headed ant on nearly all endemic insects in plant communities below about 610 m (2,000 ft) elevation where ants were abundant. He described the dramatic boundaries between areas occupied by this species and ant-free areas, especially with regard to native beetle populations. Gagné (1979) stated that at elevations above about 760 m (2,500 ft), spiders and true bugs predominated on Hawai'i Island transects, whereas below this, the arthropod biomass is lower and comprised mostly of ants; at sea level, large roaches and their parasites were abundant. Unfortunately, big-headed ants extirpated many lowland arthropods before any detailed knowledge of the fauna was obtained (Zimmerman 1978). Since big-headed ants are active throughout the night and day and reach extremely high densities, a wide range of lowland pollinating insects has likely been reduced or eliminated during vulnerable life stages or rest periods.

The long-legged ant *(Anoplolepis longipes)* was introduced to Hawai'i in 1952. It currently dominates lowland areas on windward Hawai'i Island and is found up to

1,190 m (3,900 ft) in elevation in thermal areas (C. Jorgensen and H. Black, pers. comm. 1989). These ants reach extremely dense populations in some areas, are often found in trees, and seem to eat a variety of foods. They undoubtedly affect populations of arthropod pollinators of native plants that still remain in lowland areas.

The Argentine ant *(Iridomyrmex humilis)* was introduced to Hawai'i in 1940 (Zimmerman 1941). It may be the only aggressive upland ant in Hawai'i and has been found to greatly reduce many kinds of ground-dwelling arthropods in Haleakala National Park at elevations above 1,980 m (6,500 ft) (Cole et al. 1986c). Invertebrates impacted include native species of *Nesoprosopis* bee, a known pollinator of native plants, and a number of endemic moths, bees, and flies that may pollinate the Haleakalā silversword, *Argyroxiphium sandwicense* subsp. *macrocephalum*, a species that requires cross pollination for seed set (Carr et al. 1986). Powell (1984) suggested that loss of native pollinators may be responsible for low seed set of the endangered Mauna Kea silversword *(Argyroxiphium sandwicense* subsp. *sandwicense)*.

Ground-nesting yellowjackets *(Vespula pensylvanica)* became established in Hawai'i in the late 1970s (Nakahara 1980) and are known to prey on many kinds of insects. They range as high as 2,740 m (9,000 ft) elevation. Large wasp colonies may capture and consume prey items at a rate of 10,000 per day for as long as six months; in natural areas, many prey species are native invertebrates (P.V. Gambino, pers. comm.). In native high-elevation shrublands of East Maui, endemic arthropods were prominent prey taken from yellowjacket workers returning to their nests (Gambino et al. 1987). Howarth (1985) noted sharp declines in several native arthropods at the same time as yellowjacket populations were increasing in Hawai'i. Picture-wing flies in the family Drosophilidae are among those probably affected (Carson 1982). Many species of *Drosophila*, like many of the Hawaiian honeycreepers (Drepanidinae), evolved in association with endemic rain forest plants in the family Lobeliaceae (Kaneshiro and Boake 1987).

A number of generalized species of tachinid flies and ichneumonid and braconid wasp parasitoids attack various species of Lepidoptera and other insects. Wide host selection is characteristic of some parasitoids found in both Hawaii Volcanoes and Haleakala National Parks (Stone and Loope 1987). Zimmerman (1958) believed that the introduction of various parasites to control moths of economic importance and the accidental importation of other parasites exterminated countless moth species. The ichneumonid wasp *Pristomeris hawaiiensis* parasitizes 26 microlepidopterans plus other small moths, while the ichneumonid wasp *Trathala flavo-orbitalis* has a host list of more than 30 microlepidopterans and pyralids (Zimmerman 1978). In 1958, Zimmerman wrote: "It is now impossible to see the Hawaiian Lepidoptera in the natural proliferation of species and individuals of Perkins' day. Many are forever lost." Surely, many more have also disappeared in the last 30 years.

Alien invertebrates can also transmit plant diseases. One example is the alien koa psyllid *Psylla uncatoides*, which was found to transmit native koa rusts (Leeper and Beardsley 1977). Alien aphids, leafhoppers, true bugs, and mites belong to groups that are efficient vectors (Howarth 1985). Mutualistic arrangements can also be detrimental to Hawaiian vegetation. For example, alien ants tend alien homopterans for honeydew. Some of these homopterans are important consumers of native plants and require ants to protect and disperse them. Ants, in turn, might not survive or increase in numbers and distribution without food produced from these plant-sucking insects (Howarth 1985). Honeydew produced by the native mealybug *Pseudococcus nudus* on *pūkiawe (Styphelia* tameiameiae) attracts and may provide an important carbohydrate source for yellow-jackets, which primarily feed on insects including native pollinators (Gambino et al. 1987).

Large numbers of alien invertebrates such as earthworms, ants, slugs, isopods, millipedes, and snails can cause "significant changes in the nutrient cycling process even if their direct impacts are obscure" (Howarth 1985). Changes in composition and structure of plant communities can result. Hybridization of aliens with close native relatives, can result in eventual extinction of species endemic to Hawai'i (Wells et al. 1983).

Alien Plants

Of all the disruptive forces present today in the Hawaiian Islands, feral ungulates and invasive alien plants are the two most serious threats to natural areas that have so far escaped destruction by development, agriculture, or grazing. The native flora of the Hawaiian Islands is comprised of nearly 1,000 species, 89% of them endemic (Wagner et al., in press). By contrast, nonnative plants, introduced into the Hawaiian Islands either accidentally or intentionally for human use, number approximately 4,600 (St. John 1973). More than 800 alien plant species have become naturalized (i.e., reproducing and establishing themselves without human assistance), amounting to 47% of Hawai'i's flowering plant flora (Wagner et al., in press). Of these naturalized aliens, Smith (1985) recognized 86 as serious problems in native ecosystems. He suggested that 28 of these plant pests were capable of invading undisturbed native systems.

Alien plant introductions to Hawai'i began with the arrival of the Polynesians, who brought with them about 32 species (Nagata 1985). Most of these were cultivars, and less than 25 types escaped cultivation and became naturalized (Smith 1985). Some have had such a long history in the Hawaiian Islands and have become so much a part of lowland forests that they were considered native by many earlier writers. These include *kukui (Aleurites moluccana)*, mountain apple or *'ōhi'a 'ai (Syzygium malaccense)*, and wild yam *(Dioscorea pentaphylla)* (Handy and Handy 1972). A few additional species previously thought to be either indigenous or later European introductions may actually be accidental Polynesian introductions (St. John 1978; Wester, in press).

With the arrival of Capt. Cook in 1778 and the subsequent influx of ships from Europe and North America, plant introductions accelerated. The first recorded western introductions into Hawai'i were pumpkins, melons, and onions *(Cucurbita pepo, Cucumis melo, Allium cepa)*, planted during Cook's visit. In the 60 years following Cook's voyage, at least 111 nonnative plant species were introduced (Nagata 1985). These were primarily fruit trees, vegetables, and ornamentals, many of which remained in cultivation and did not become naturalized. Notable exceptions are guava *(Psidium guajava)*, strawberry guava *(P. cattleianum)*, and *koa haole (Leucaena leucocephala)*, which became major invaders of lowland ecosystems. Most of the alien plant species threatening the integrity of native systems today are introductions of the 20th century. Wester (in press) noted that the rate of plant introduction increased sharply at the end of the 19th century, and about five species per year have naturalized during the 20th century.

The effects of alien species on native species and ecosystems are not well under-stood or documented (Loope, in press), although it is clear that their impact has been major in Hawai'i. Many of the component species of island biota are thought to exhibit "reduced aggressiveness or increased vulnerability to extinction," and the characteris-tic low number of species per unit area in island ecosystems, as well as "certain genetic properties" of island organisms, may contribute to this vulnerability (Loope and Mueller-Dombois 1989). Because of limited ranges, consequent small population sizes, and habitat specialization, endemic Hawaiian species are vulnerable to disturbance (Simon 1987). Alien plants (and animals) have contributed to the extinction of native species in the lowlands of Hawai'i and have been a "primary cause" of such extinction in upland habitats (Vitousek et al. 1987).

The most-often cited effects of alien plants on native plant species are competition and displacement; competition may be for water or nutrients, or it may involve allelo-pathy (chemical inhibition of other plants) (Smith 1985). Alien plants may displace natives by preventing their reproduction, usually by shading and taking up available sites for seedling establishment (Vitousek et al. 1987). Alien plant invasions may also alter entire ecosystems by forming monotypic stands, changing fire characteristics of native communities, altering soil-water regimes, changing nutrient cycling, or en-couraging other nonnative organisms (Smith 1985; Vitousek et al. 1987). Hawai'i is now recognized as an important "natural laboratory" to study interactions among native and alien species and the invasions of aliens in natural ecosystems (Simon 1987; Vitousek et al. 1987).

The following discussion will cover only a few of the most widespread and disruptive of alien plant invaders: those primarily found in wet zones, then those characteristic of dry and mesic habitats. For a more inclusive discussion of disruptive alien plants in Hawai'i, see Smith (1985).

Invaders of Wet Habitats

Banana Poka and Other Passiflora. A vine or liana in the passionflower family, ba-nana poka *(Passiflora mollissima)* was introduced to the island of Hawai'i before 1921, when it was observed growing wild at Pu'uwa'awa'a Ranch in the North Kona Dis-trict (LaRosa 1984). Native to the Andes (St. John 1973), the vine was probably brought to Hawai'i as an ornamental because of its showy, pink flowers. Banana poka was apparently planted at most of the sites that currently have heavy infestations: Hualālai; Keanakolu/Laupāhoehoe on Mauna Kea; Wright Road/'Ola'a on Mauna Loa; Kula on Maui; and Kōke'e, Kaua'i (Warshauer et al. 1983; LaRosa, in press). As of 1981, ba-nana poka was distributed over 520 km^2 (200 mi^2) in wet and mesic forests on Hawai'i and Kaua'i, of which 190 km^2 (73 mi^2) represented continuous distribution. On Maui, banana poka is found on 200 ha (500 a) at upper Waiakoa, near Kula, where it was previously thought to have been eradicated by State workers (Smith 1985). The relatively small Maui population has undergone dramatic expansion since 1971 (LaRosa, in press) but has recently received attention from volunteer groups, and there is hope that complete eradication is still possible with the aid of special fund-ing to the Department of Land and Natural Resources from 1989 to 1991 (B. Gagné and L. Loope, pers. comm. 1989).

Most of the poka-infested acreage occurs on the island of Hawai'i, with three main areas of concentration and several isolated populations. The oldest infestation is in the North Kona District in mesic forests of the slopes of Hualālai and Mauna Loa, where banana poka vines are scattered rather than continuously distributed over most of its range. Warshauer et al. (1983) interpreted the distribution pattern to mean that banana poka is still spreading into adjacent suitable habitat.

The second-oldest infestation is in *koa/'ōhi'a (Acacia koa/Metrosideros polymorpha)* forest of the windward slopes of Mauna Kea and is called the "most advanced or severe" infestation surveyed. When banana poka distribution in 1983 was compared with that found in a survey ten years earlier, the size of this population seemed relatively stable.

The third major infestation occurs in montane wet *'ōhi'a/hāpu'u (Metrosideros polymorpha/Cibotium* spp.) forest in 'Ola'a Tract of Hawaii Volcanoes National Park, where the vine increased in cover 18-fold between 1971 and 1981 (Warshauer et al. 1983). Warshauer and others blamed the banana poka extension and intensification on a major wind storm in 1980, which opened up the tree fern (*Cibotium* spp.) canopy in 'Ola'a. LaRosa (in press) considered the 'Ola'a population to be the youngest and most rapidly expanding infestation on the Island. In a recent (1988) ground survey of 'Ola'a in Hawaii Volcanoes National Park, banana poka was found to be widely distributed, with dense cover in the western third of the Tract (S.J. Anderson et al., unpubl. data).

Banana poka is extremely detrimental to native forests and can change their structure and species composition (Warshauer et al. 1983). Heavy loads of vines may cause damage or death to native trees, and when branches break or trees fall, the tree canopy is opened, changing understory conditions. Dense cover of banana poka in forest canopies reduces the sunlight reaching trees and may decrease photosynthesis (LaRosa, in press). Much of the Hawai'i Island infestation is in koa forests; here koa reproduction is affected, as banana poka outgrows koa seedlings and saplings and competes with them (Scowcroft and Nelson 1976). Bird populations of infested forests may be adversely affected by the disturbance brought about by banana poka, particularly those endangered bird species whose primary habitat is the koa/'ōhi'a forest (Warshauer et al. 1983; Scott et al. 1986). One native bird, the *'i'iwi (Vestiaria coccinea)*, has become "associated with *Passiflora*" and has been observed feeding on banana poka nectar (Scott et. al 1986).

Banana poka is spread by alien fruit-eating birds such as the kalij pheasant *(Lophura leucomelana)* (Lewin and Lewin 1984), as well as by the feral pig *(Sus scrofa)* (Warshauer et al. 1983). Feral pigs open up forest understory, create bare disturbed areas in which weeds can establish, and topple tree ferns, encouraging the sun-loving banana poka. LaRosa (1984) reported that banana poka invaded and increased in density in 'Ola'a forest plots after the tree fern cover was artificially decreased or removed. Dispersal of this plant by alien animals, as well as natural perturbations such as windstorms and treefalls, make banana poka difficult to control. Previous attempts to reduce banana poka with herbicides on heavily infested State lands were largely unsuccessful (LaRosa 1984). Efforts are currently under way to control the plant in certain relatively intact portions of Hawaii Volcanoes National Park (Cuddihy et al. 1988), and preliminary research into mechanical and herbicidal control methods has been completed (Santos et al. 1989a). Biological control research is ongoing, and one potential control agent, an iridescent blue moth *(Cyanotricha necyria)*, whose larvae defoliate poka leaves, has already been released on Hawai'i Island (Markin et al., in press).

If left unchecked, the future distribution of banana poka could be far greater than it is today. Jacobi and Warshauer (in press) modeled potential banana poka range on the island of Hawai'i based on rainfall and elevation and concluded that the species could become established over a vast region between 500 and 2,500 m (1,640-8,200 ft) elevation on the slopes of all five of the Island's volcanoes.

Other species of *Passiflora* are also pests in natural areas of the Hawaiian Islands, although they currently have a more limited distribution than banana poka. Sweet granadilla *(P. ligularis)* has habits similar to banana poka and is a problem on Hualālai, Hawai'i Island; significant populations also exist in the South Kona and Puna Districts. *Huehue-haole (P. suberosa)*, a vine cabable of smothering small trees and understory plants, has infested drier areas, such as the Wai'anae Mountains of O'ahu and leeward East Maui (Smith 1985).

Guavas: Strawberry Guava, Common Guava. An early European introduction native to Brazil (St. John 1973), strawberry guava or *waiawī (Psidium cattleianum)* has been in the Hawaiian Islands since 1825, when several species of guava and other cultivated fruit plants arrived on the British ship *Blonde* (Nagata 1985). Today strawberry guava occurs on all the main Hawaiian Islands except Ni'ihau and Kaho'olawe, generally in wet and mesic lowland regions between 150 and 1,300 m (495-4,260 ft) elevation (Smith 1985; Wagner et al., in press). During the U.S. Fish and Wildlife Service Forest Bird Survey, strawberry guava was found at 8% of the stations sampled above 500 m (1,650 ft) elevation on the island of Hawai'i, suggesting that it is a very widespread alien species (Jacobi and Warshauer, in press).

In Hawaii Volcanoes National Park, strawberry guava is found from near sea level to above 1,220 m (4,000 ft) elevation and is acknowledged as a disruptive invader of native systems (Tunison et al., in press). In a recent (1988-89) survey, strawberry guava was found to be distributed at low densities over nearly half of 'Ola'a Tract and was even more widespread and abundant in *'ōhi'a (Metrosideros polymorpha)* forests on the Park's East Rift (S.J. Anderson et al., unpubl. data). Strawberry guava is recognized as the most serious plant pest of the Kīpahulu District of Haleakala National Park, Maui (National Park Service 1986b), where it has invaded *koa (Acacia koa)* and 'ōhi'a rain forests up to an elevation of 1,140 m (3,740 ft) (Yoshinaga 1980) and forms very dense stands in lower-elevation forests. Strawberry guava potentially threatens numerous low- to middle-elevation rain forest plant species in Haleakala National Park with extirpation through displacement. Species particularly threatened in this way include *hame (Antidesma platyphyllum)*, *po'olā (Claoxylon sandwicense)*, *kuenui (Cyanea grimesiana)*, *'ohe (Joinvillea ascendens)*, *'aiea (Nothocestrum longifolium)*, and *nuku'i'iwi (Strongylodon ruber)* (Loope et al., in press b).

Smith (1985) considered strawberry guava to be the "worst pest in Hawai'i's rain forests." In favorable habitats, the tree is capable of forming nearly monotypic stands, excluding almost all native ground cover and other understory species. In forests with serious strawberry guava infestations, most native plants do not regenerate for several reasons: very dense shade, digging by feral pigs *(Sus scrofa)* that are attracted by strawberry guava fruit, and allelopathic effects of strawberry guava leaf litter (Smith 1985; Wagner et al., in press). Even where strawberry guava does not form dense stands, its presence encourages higher pig populations and may adversely affect rare plant species with limited distributions. The seeds of strawberry guava are known to be spread by both feral pigs and frugivorous birds. L.F. Huenneke and P.M. Vitousek (unpubl. data) found that strawberry guava has very high rates of fruit production and seed germination; they also attributed some of the plant's success in native forests to its ability to readily produce sturdy, fast-growing suckers, which are less injured by falling branches than are more delicate native tree seedlings.

Strawberry guava is considered to be a pest by cattle ranchers because of its ability to form dense stands and crowd out forage plants (Hosaka and Thistle 1954). The species is currently recognized as a forest pest by State foresters (Hostetler 1970), but in the past (1928-52) more than 26,000 strawberry guava trees were planted in Oʻahu and Kauaʻi forest reserves (Skolmen 1979).

Control of strawberry guava may be possible with herbicides applied to cut stumps or frilled trunks (Gardner 1980; Santos et al., 1989b), at least in limited areas with important natural resources (Tunison, in press a). However, such control is extremely labor intensive and cannot be applied to dense and extensive infestations. The economic value of common guava (*Psidium guajava*) was previously thought to preclude biological control research (Gardner and Davis 1982); however, several potential insect control agents were recently located on native stands of strawberry guava during exploration in Brazil, suggesting the possibility of future successful biocontrol (Hodges 1988). Control of this pest is critical; Jacobi and Warshauer (in press) predicted that the plant may be able to invade nearly half the area surveyed by the U.S. Fish and Wildlife Service Forest Bird Survey on the island of Hawaiʻi, by "filling in" areas among current infestations below 1,300 m (4,260 ft) elevation.

Common guava, another early introduction from tropical America, is also considered a pest by many. Like strawberry guava, it can form dense thickets in wet and mesic habitats of the lowlands. However, unlike its congener, common guava is usually found in previously disturbed areas and is abundant primarily below 500 m (1,650 ft) elevation (Smith 1985; Wagner et al., in press). This is the plant that Isabella Bird (1966) found so common on the sides of ridges and in the understory of forests she traversed on Hawaiʻi Island in 1873. Other trees of the myrtle family have also become pests and lowland forest invaders in the Hawaiian Islands, most notably two species in the genus *Syzygium* (formerly *Eugenia*): Java plum *(S. cumini)* and roseapple *(S. jambos)*.

Melastomes: Clidemia, Tibouchina, Others. The melastome family (Melastomataceae) is not represented in the native flora of Hawaiʻi, but there are 14 naturalized species in 11 genera, and at least five have become disruptive invaders of native ecosystems (Wagner et al., in press). One of the worst is *Clidemia hirta* or Koster's curse, which has caused great concern because of its rapid spread in the Hawaiian Islands. Native to tropical Central and South America, *Clidemia* has become widely naturalized in other tropical regions. In Fiji, *Clidemia* was introduced in the late 19th century and became a serious pest in pasturelands and rubber and coconut plantations (Wester and Wood 1977).

Clidemia was first reported in the Hawaiian Islands in 1941 on Oʻahu near Poamoho; it may have been originally planted in the nearby Wahiawā Botanic Gardens. Within the Koʻolau Mountains, *Clidemia* spread in the 1940s (Smith, in press), and by the early 1960s the pest had expanded from Tantalus north to Kawailoa Ridge (Wester and Wood 1977) but was not known to be on the other Hawaiian Islands (Plucknett and Stone 1961). By 1970, *Clidemia* had spread to the Waiʻanae Mountains, where it is now widespread in the Honouliuli Forest Reserve (Smith, in press). The rapidity of the spread and intensification of *Clidemia* on Oʻahu is remarkable: between 1977 and

1988, the species tripled its range from 31,350 ha (77,430 a) (Wester and Wood 1977) to more than 100,000 ha (247,000 a) (Smith, in press).

Clidemia spread to five other Hawaiian Islands in the 1970s and 1980s. On Kaua'i, there are now at least five small populations, totalling about 40 ha (100 a). Moloka'i currently has an infestation totalling more than 6,000 ha (14,820 a), primarily in Wailau and Pelekunu Valleys of the North Shore; much more available habitat on Moloka'i may be invaded in the next few years. East Maui was invaded in the mid 1970s, and by the early 1980s *Clidemia* was found on West Maui (Smith, in press). At present, there are two established populations of *Clidemia* on East Maui: at Makapipi (30 ha or 75 a), and at 'O'opuola and adjacent stream drainages (1,200 ha or 3,000 a) (Medeiros et al. 1989). Recently (1988), two *Clidemia* plants were discovered on the island of Lāna'i (Smith 1988). The island of Hawai'i has at least four populations on windward slopes, all reported between 1972 and 1983 (Smith, in press).

The effects of *Clidemia* on native vegetation are devastating. The shrub replaces native plants of the forest understory, and almost nothing except a few hardy mosses can grow beneath its shade. As *Clidemia* replaces the multilayered understory of native forests, their value as watersheds may be greatly reduced (Smith, in press). Even the hardy and aggressive native matted fern *uluhe (Dicranopteris linearis)*, a dominant cover on many slopes and in lowland forests, can be replaced by *Clidemia* in shaded areas (Wester and Wood 1977). Native insects and birds are probably lost along with the native vegetation in invaded areas. While not resistant to fires (Smith 1985), *Clidemia* can take advantage of such disturbances. Two years after a fire in Kawailoa Forest Reserve on O'ahu, *Clidemia* had become the dominant cover and achieved heights greater than 2 m (7 ft) (Wester and Wood 1977). *Clidemia* is considered a pest by cattle ranchers, as it has no forage value (Hosaka and Thistle 1954).

Clearly, the plant is easily dispersed. Birds are known to eat the berries (Hosaka and Thistle 1954), and the ubiquitous Japanese white-eye *(Zosterops japonica)* may be a prime vector (Wester and Wood 1977). Hikers, hunters, and marijuana *(Cannabis sativa)* growers are responsible for accidentally dispersing *Clidemia*; feral pigs *(Sus scrofa)* both spread the weed and encourage its intensification through digging and soil disturbance (Smith, in press).

There is hope that biological control may help reduce the impacts of *Clidemia*. Nakahara et al. (in press) reported that of 14 insects from Trinidad recently evaluated as *Clidemia* control agents, three to five may be released in Hawai'i in the near future. A fungus from Panama *(Colletotrichum gloeosporoides)*, which attacks the leaves of *Clidemia*, was introduced into Hawai'i in 1986 (Anon. 1986; TenBruggencate 1986b). A thrips *(Liothrips urichi)*, introduced earlier (1953) to combat *Clidemia*, has been effective on ranchlands but not in forests (Reimer 1985). Other insects released to control the weed have been even less successful (Nakahara et al., in press). If biological control is not effective on this species, the future of remaining low-elevation wet and mesic forests is grim. Seeds in the soil may remain viable for more than four years. Mechanical control is very time consuming, and in rain forests, even the leaves of pulled plants may form new roots and reestablish (Smith, in press).

Three species of *Tibouchina* have become naturalized in the Hawaiian Islands, but only two currently represent a threat to native vegetation. Glorybush or princess flower *(T. urvilleana)*, native to Brazil, was introduced to Hawai'i in 1910 as an ornamental. A shrub with showy, purple flowers, it is still cultivated around residences and has become naturalized in wet areas on Kaua'i, O'ahu, Maui, and Hawai'i (Wagner et al., in press). In 1936-38, more than 1,000 glorybush were actually planted in forest reserves on O'ahu and Kaua'i (Skolmen 1979). Like other members of the melastome family, glorybush can form dense thickets and crowd or shade out native plants; the shrubs may achieve heights greater than 4 m (13 ft) (Smith 1985). In some areas, glorybush may penetrate less readily into forests than other melastomes (Plucknett and Stone 1961).

Glorybush spread rapidly from plantings in the Kōke'e area of Kaua'i and is considered a threat to native plants there (Hadley 1966). In Hawaii Volcanoes National Park, the species is localized in wet 'ōhi'a *(Metrosideros polymorpha)* forests around the rim of Kīlauea Crater, where it has proven difficult to eradicate (J.T. Tunison, pers. comm. 1988). Although it seems not to produce viable seeds (Wagner et al., in press), cut, broken, or decumbent branches of glorybush readily root in humid montane climates. A relatively effective cut-stump herbicide treatment for glorybush is known, but the shrub's tangled growth form makes such control extremely laborious (Santos et al. 1986). A moth introduced as a biological control agent *(Selca brunella)* has become established on glorybush (Gardner and Davis 1982) but is apparently not seriously affecting the species (Smith 1985).

Tibouchina herbacea, native to several South American countries, became established in the late 1970s on Hawai'i Island and was noted on East Maui in 1982 and West Maui in 1986 (Wagner et al., in press). First collected on Hawai'i Island near the Saddle Road, this shrub has been rapidly spreading in wet and mesic forests and has recently been reported from Pu'u o 'Umi Natural Area Reserve (Smith 1988), Kahauale'a Natural Area Reserve (S. Perlman, pers. comm. 1989), and remote forests of the East Rift of Kīlauea in Hawaii Volcanoes National Park (S.J. Anderson et al., unpubl. data). The disruptive capabilities of this species are not fully known, but *T. herbacea* density seems to be intensifying in forests with 'ōhi'a dieback.

Other members of the melastome family are also potential threats to native forests. One recently naturalized species is *Miconia calvescens* or velvet tree, native to tropical America (Wagner et al., in press). *Miconia* has been planted as an ornamental in yards and botanic gardens on Hawai'i Island (Davis 1987), Maui (Smith 1988), and O'ahu (B. Gagné, pers. comm. 1989). On the island of Hawai'i at least, *Miconia* has escaped cultivation and become established near Onomea, Hilo, and Pāhoa (Davis 1987). While the current infestations in Hawai'i are small, *Miconia* has the potential to be seriously disruptive; after introduction to Tahiti and Moorea, *Miconia* rapidly invaded native forests up to 1,220 m (4,000 ft) elevation and became an understory dominant (B. Gagné, pers. comm. 1989).

Another member of the melastome family, *Oxyspora paniculata*, is a pink-flowered shrub that escaped from cultivation on Tantalus and became established on O'ahu before 1954 (Wagner et al., in press). Although it was considered non-invasive for more than 20 years, recently *Oxyspora* has been spreading eastward along the summits of the Ko'olau Mountains and has become the dominant plant along some mountain stream banks (Obata 1985a). Because of this "population explosion," Obata raised the specter of *Oxyspora* as another *Clidemia*. That *Oxyspora* is continuing its spread on O'ahu is indicated by recent sightings on Wa'ahila Ridge (Smith 1988) and Mānoa Valley (C.W. Smith, pers. comm. 1989).

Rubus: Yellow Himalayan Raspberry, Blackberry, Others. At least six species of *Rubus* have become naturalized in the Hawaiian Islands. Three of these species now infest wet and mesic ecosystems on several islands, and one is also a problem in subalpine grasslands. The three remaining species occur over smaller and more limited areas, but they have potential to invade natural areas (Wagner et al., in press).

Rubus ellipticus, or yellow Himalayan raspberry, is a recent introduction into the State (c. 1961) and is currently restricted to the island of Hawai'i. Perhaps introduced as an ornamental or for its insipid, yellow fruit, the raspberry is native to India (Wagner et al., in press). There is a major infestation in the Volcano area and Hawaii Volcanoes National Park, but the plant may be seen between 700 and 1,700 m (2,300-5,580 ft) in wet habitats elsewhere on the Island (Smith 1985). Like other *Rubus* species, this raspberry may form large, tangled thickets and will grow in wet forests, pastures, open areas, and roadsides. Unlike Florida blackberry *(R. argutus)*, yellow Himalayan raspberry is capable of growing to enormous size even in the dense shade of closed-canopy rain forests. In a recent (1988) survey of 'Ola'a Tract in Hawaii Volcanoes National Park, yellow Himalayan raspberry was found to be distributed over more than half of the 3,770-ha (9,300-a) Tract (S.J. Anderson et al., unpubl. data). This indicates a dramatic expansion since an earlier survey of 'Ola'a Tract by Jacobi and Warshauer (1975), when the species was recorded primarily in a 100-ha (250-a) *kīpuka* (island of older vegetation) of *koa (Acacia koa)* and along nearby roadsides.

During the U.S. Fish and Wildlife Service Forest Bird Survey of 1976-81, yellow Himalayan raspberry was found on only 0.1% of the stations sampled (Jacobi and Warshauer, in press), but the authors considered this *Rubus* to be in "an incipient stage of invasion" and predicted that it had the potential to invade vast regions on both the windward and leeward slopes of Hawai'i Island. *Rubus ellipticus* has been reported to be capable of nitrogen fixation on the island of Java (Becking 1979), although this has not been demonstrated in Hawai'i. Biological control work has not been initiated on this species (Smith 1985), but research has been conducted on potential herbicidal control methods (Santos et al. 1986) with some success.

Another serious invader is prickly Florida blackberry (*R. argutus*, formerly known in Hawai'i as *R. penetrans*), introduced from the eastern or central United States perhaps as early as 1894 (Haselwood and Motter 1976). Recognized as a pest in the 1930s (Doerr 1931), by 1962 blackberry was distributed over nearly 18,000 ha (44,460 a) on four Islands (Santos et al. 1986). It now occurs on five of the main Islands between 1,000 and 2,300 m (3,280-7,540 ft) elevation (Smith 1985) and is considered a noxious weed by the Hawaii State Department of Agriculture (1978).

Easily spread by fruit-eating birds and capable of vigorous vegetative reproduction, this plant can form dense, impenetrable thickets in open vegetation (Smith 1985). Montane mesic forests of Hawaii Volcanoes National Park have been heavily invaded, and Florida blackberry has also been found in nearly intact rain forest and subalpine grasslands of Haleakala National Park on Maui (Loope et al., in press b). On Kaua'i, blackberry has invaded the rain forests of Kōke'e, where it appears to be competing with the native plant understory and "smothering native ferns" (Hadley 1966).

Rubus rosifolius, locally known as thimbleberry, is perhaps the most widespread of all the introduced species of *Rubus*. This Asian plant was introduced to the island of Hawai'i from Jamaica in the early 1880s, and despite regulations to prohibit interisland transport (Degener 1936), thimbleberry became established in wet forests on all the main Hawaiian Islands (except Ni'ihau) from near sea level to above 1,730 m (5,670 ft) elevation (Wagner et al., in press). A relatively small shrub, thimbleberry seems to be much less aggressive than other members of the genus, although it may be very abundant locally, particularly in areas disturbed by feral pigs *(Sus scrofa)*.

Other naturalized species of *Rubus* currently have more limited distributions in Hawai'i. A relatively recent infestation of *Rubus glaucus* in 'Ola'a Tract of Hawaii Volcanoes National Park is growing in dense *'ōhi'a*/tree fern *(Metrosideros/Cibotium)* rain forest. This species was apparently planted at a nearby agricultural experiment station and then abandoned as a potential crop, without destruction of the plants (Smith 1985). Two other cultivated raspberries, *Rubus niveus* and *R. sieboldii*, native to Asia, were introduced to one or more of the Islands for their fruit; these potential pests have escaped cultivation and become naturalized (Smith 1985; Wagner et al., in press). Bryan (1954) reported that seeds of *R. niveus* (then called *R. albescens*) were imported from Florida in 1952 and were germinated and distributed to the public by the territorial Division of Forestry; this *Rubus* was considered to be safe to distribute because plants did not "send out adventurous suckers from the root." Despite the history of *Rubus* infestations in the Hawaiian Islands, nearly 100 *Rubus* plants, unidentified to species, were planted in four forest reserves between 1938 and 1954 (Skolmen 1979).

Gingers: White, Yellow, and Kahili Ginger. Two members of the ginger family were brought to Hawai'i by the Polynesians: *'ōlena (Curcuma longa)* and shampoo ginger or *'awapuhi (Zingiber zerumbet)*. Both have become naturalized, usually in previously disturbed habitats. Among the many ornamental gingers that have been introduced to Hawai'i in the last 200 years, six have become naturalized (Wagner et al., in press), and three are problems in native ecosystems.

White ginger *(Hedychium coronarium)* and the very similar yellow ginger *(H. flavescens)* were both introduced in the late 1800s, probably by Chinese immigrants (Degener 1934). White ginger is from China and the Himalayas, and its yellow-flowered relative is native to northern India. Both species escaped cultivation and are now established in lowland wet and mesic forests on most of the main Hawaiian Islands (Wagner et al., in press). In suitable habitats, these large, vigorous herbs are capable of forming a very dense ground cover that excludes all other growth. Although most common along roadsides, trails, and other disturbed areas, they have also invaded streambeds in many natural areas. A nearly pure stand of ginger was recently noted in streambeds of The Nature Conservancy's Kamakou Preserve on Moloka'i (Cuddihy et al. 1982b). Fortunately, neither white nor yellow ginger is a heavy producer of conspicuous fruit or seed, so vegetative reproduction predominates (Smith 1985).

Kahili ginger *(H. gardnerianum)* is a more serious threat to native forests because of its ability to produce abundant fruit, with orange seeds attractive to many alien birds. Native to the Himalayas, this ginger was introduced to Hawai'i sometime before 1940 (Wagner et al., in press) and was first collected in Hawaii Volcanoes National Park in 1943 (Fagerlund 1943). At first, kahili ginger was not perceived as a threat to native forests of the Park because it remained restricted to gardens in the Park housing area, at least until 1947 (Fagerlund 1947). However, by the mid-1980s, kahili ginger had become established at high densities in the Kīlauea area more than a mile from original plantings, and it was found at low densities in Kīpuka Puaulu and 'Ola'a Tract, more than 5 km (3 mi) from the housing area. Additional isolated populations occur on Kīlauea's East Rift, more than 10 km (6 mi) from Park housing (S.J. Anderson et al., unpubl. data); these populations probably represent dispersal from other plantings.

Major infestations also occur at Kōke'e, Kaua'i, and windward East Maui (Smith 1985). In Kīpahulu Valley on Maui, kahili ginger has been recently (1988) sighted as high as 1,220 m (4,000 ft) elevation, in nearly intact rain forest (Loope et al., in press b). This ginger was not found in previous surveys for weeds of the middle and upper Valley (Yoshinaga 1980), and it appears to be rapidly invading from more disturbed habitats at lower elevations. Kahili ginger is particularly insidious in native rain forest because, unlike most alien plants, it can form very dense stands even under the heavy shade of intact canopies of native trees and tree ferns. Kahili ginger typically achieves heights greater than 1 m (3 ft) and can displace all other understory plants in its vicinity (Smith 1985).

Although a number of diseases and insects are known to attack gingers, biological control is thought to be unlikely because of opposition from horticulturists and commercial ginger producers (Gardner and Davis 1982; Smith 1985). Herbicidal control methods have been investigated but are not completely effective (Santos et al. 1986). Mechanical removal of ginger by digging up rhizome masses has been accomplished in small portions of Hawaii Volcanoes National Park, but this method of control is extremely time consuming and labor intensive (Cuddihy et al. 1988; Tunison, in press a).

Australian Tree Fern. Ferns are not generally considered to be disruptive to native systems in Hawai'i; only two species, *Athyriopsis japonica* and *Blechnum occidentale*, are listed by Smith (1985) as potential pests. Australian tree fern (*Cyathea australis*) is a recent introduction into Hawai'i that has been planted in private and hotel gardens at low elevations on O'ahu, Maui, and Hawai'i. It is naturalized on O'ahu (C.W. Smith, pers. comm. 1988), and in 1988 naturalized populations of this fern were found in Kīpahulu Valley, Haleakala National Park, at elevations as high as 1,040 m (3,400 ft) (Anderson et al., in press). Some of the individuals seen were taller than 4 m (13 ft), which undoubtedly represents the growth of many years, even though this alien plant was not listed on previous plant surveys of the Valley (Lamoureux 1967; Yoshinaga 1980). The future of this infestation and possible competition between this fern and native tree ferns (*Cibotium* spp.) in the Valley are unknown, but monitoring and control have been recommended (Loope et al., in press b). Ironically, another recently introduced *Cyathea (C. cooperi)*, is being promoted for commercial propagation in West Hawai'i with the reasoning that production of the faster-growing *Cyathea* will decrease exploitation of native tree ferns and reduce destruction of native forests (Anon. 1989 b).

Grasses: Hilo Grass, Palmgrass, Meadow Ricegrass. Alien grasses have long been perceived as threats to Hawaiian forests. Three species, Hilo grass, palmgrass, and meadow ricegrass *(Paspalum conjugatum, Setaria palmifolia,* and *Ehrharta stipoides,* syn. *Microlaena),* are especially important in wet habitats in Hawai'i. Two additional species, dallis grass and Vasey grass *(Paspalum dilatatum* and *P. urvillei)* are widespread in disturbed areas on the islands of Maui and Hawai'i and have invaded some rain forests and montane mesic communities.

One of the earliest recognized invaders of wet forests was Hilo grass *(Paspalum conjugatum)*. Native to tropical America, this grass was first noted near Hilo in 1840, hence the common name in Hawai'i. Hilo grass spread rapidly and is now distributed in wet habitats from near sea level to 2,000 m (6,560 ft) elevation (Smith 1985) on all the main Islands except Ni'ihau and Kaho'olawe (Wagner et al., in press). In the early 1900s, Hilo grass was recognized as a serious threat to watersheds and forest reserves, and its presence was considered an impediment to planting and reforestation (Giffard 1918). Able to form a dense cover even on poor soils (Smith 1985), Hilo grass was blamed for preventing the reproduction and spread of native ferns and other plants and for converting forest to open grassland (Lyon 1921; Merrill 1941). Lyon (1922) described how Hilo grass could overtop and eliminate herbaceous undergrowth and smother native tree seedlings. He observed that 'ōhi'a (*Metrosideros polymorpha*) seeds scattered over Hilo grass never became established as seedlings.

Hilo grass remains a problem in forest reserves and national parks in Hawai'i, particularly at lower elevations. In Kīpahulu Valley of Haleakala National Park, Hilo grass is abundant below 1,000 m (3,280 ft) elevation and has apparently been increasing in cover over the last 20 years, perhaps because of disturbance from feral pigs (*Sus scrofa*) (Yoshinaga 1980). Between 1980 and 1986, Hilo grass expanded its range up-slope in Kīpahulu Valley by more than 180 m (600 ft) elevation and by 1986 was found to dominate ground cover below 1,140 m (3,740 ft) (Anderson et al., in press). Re-examination of an exclosure at 975 m (3,200 ft) elevation in Kīpahulu Valley (after 10 years of protection and drastic reduction of the feral pig population) suggests a small decline in Hilo grass, recovery of fragile native ground ferns, and reproduction of native species (A.C. Medeiros and L.L. Loope, pers. comm.). The future recovery of forests heavily invaded by Hilo grass appears jeopardized, because it is not certain that infested areas will ever completely recover, even in the absence of disturbance from feral animals. Biological control research is not planned for this species (Smith 1985).

Palmgrass (*Setaria palmifolia*), introduced from tropical Asia before 1903, has become naturalized on O'ahu, Lāna'i, Maui, and Hawai'i (Wagner et al., in press). Currently, this tall (to 2 m or 7 ft) perennial grass may be found in wet forests to an elevation of 2,000 m (6,560 ft), with major infestations in the Ko'olau Forest Reserve on Maui, in four reserves on windward Hawai'i, and in the 'Ola'a Tract of Hawaii Volcanoes National Park (Smith 1985). In a recent survey of 'Ola'a (1988-89), approximately one quarter of the Tract was found to have a dense ground cover (>50%) of palmgrass (S.J. Anderson et al., unpubl. data). By contrast, when Jacobi and Warshauer (1975) surveyed the 'Ola'a Tract 14 years ago, they found the species only along trails and in very few disturbed localities. When this large grass invades forests, it may form very dense stands and shade out all other herbaceous vegetation (Smith 1985). Fosberg (1948b) noted that palmgrass, in less than 20 years, spread from a single colony to become the dominant herbaceous plant along a trail in Mānoa Valley on O'ahu. He also observed that once established, palmgrass eradication is very difficult.

A third grass with disturbing potential in both wet and mesic habitats is meadow ricegrass (*Ehrharta stipoides*, syn. *Microlaena*), a native of Australia, New Zealand, and the Philippines; it arrived in Hawai'i before 1916 and is now found on O'ahu, Maui, and Hawai'i (Wagner et al., in press). Meadow ricegrass is capable of rapidly invading disturbed sites and forming dense stands. It has become established over large areas in the Puna, Ka'ū, and South Kona Districts of the island of Hawai'i (Smith 1985) and has also invaded important endangered bird habitat at Hakalau Forest National Wildlife Refuge (Stone et al., in press). While the effects of meadow ricegrass have not been fully assessed, it is capable of growing in the moderate shade of many mesic forests, and its dense ground cover may be capable of inhibiting the reproduction of native forest plants. The disappearance of colonies of the legume *Vicia menziesii*, Hawai'i's first listed endangered plant species, has been associated with meadow rice-grass invasions subsequent to disturbance from clearing and logging activities (Clarke et al. 1983).

Invaders of Dry and Mesic Habitats. Most of the dry lowlands of the Hawaiian Islands are completely dominated by alien plants, particularly shrubs and grasses (Smith 1985). A few of the worst invaders of remaining native plant habitats in dry and mesic regions are discussed below.

<u>**Trees: Firetree, Silk Oak**</u>. Firetree, or faya tree *(Myrica faya)*, has attracted a great deal of attention and concern for its recent explosive increase on several Islands and its ability to invade nearly intact native ecosystems. Native to the Azores, Madeira, and the Canary Islands, firetree occurs there in laurel *(Laurus azorica)* forests from sea level to 610 m (2,000 ft) elevation and is also found in pastures within its native range (Hodges and Gardner 1985).

Firetree was introduced to Hawai'i before 1900, probably by Portuguese immigrants as an ornamental, a source of fruit for wine-making (Wagner et al., in press), or for firewood (Hodges and Gardner 1985). Subsequently, the Hawaii Sugar Planters' Association obtained seeds from a Portuguese farmer on Hawai'i (Whiteaker and Gardner 1985), and the tree was planted in ten forest reserves on Kaua'i, O'ahu, and Hawai'i, primarily in 1926-27 (Skolmen 1979). L.W. Bryan recorded collecting *Myrica* seed from Hawai'i and sending it to a forest supervisor on Kaua'i in 1926 (Bryan 1926a). By the mid-1980s, firetree had infested more than 34,000 ha (83,980 a) Statewide, with most of the infestation on the island of Hawai'i and smaller acreages on Maui, Lāna'i, O'ahu, and Kaua'i (Whiteaker and Gardner 1985). The Maui infestations occur primarily on ranchlands, and the O'ahu firetree distribution includes remnant 'ōhi'a *(Metrosideros polymorpha)* cloud forest and mesic shrublands of the southern Wai'anae Mountains. On Kaua'i, firetree has invaded wet and mesic montane rain forests of Kōke'e State Park. On Hawai'i Island, firetree occurs in pasturelands of the Hāmākua and Ka'ū Districts and on the slopes of Hualālai, as well as in native rain forests and open 'ōhi'a forest and shrublands in and near Hawaii Volcanoes National Park (Whiteaker and Gardner 1985).

The history of firetree in Hawaii Volcanoes National Park illustrates the rapidity of its invasion and intensification. In 1961, only one individual tree was known to occur in the Park, at Kīlauea Military Camp (Fosberg 1966), although others probably existed away from trails and roads. By 1966, approximately 90 ha (225 a) in the Park were infested with firetree, and in just 11 years (1977) this infestation increased explosively to 3,640 ha (9,000 a) (Smathers and Gardner 1979). By 1985, firetree covered more than 12,200 ha (30,130 a) of the Park (Whiteaker and Gardner 1985).

The impacts of firetree on native ecosystems are serious. Given enough time in a suitable site, firetree can form a dense, closed-canopy, nearly monospecific stand of trees 15 m (50 ft) tall, with virtually no plant cover on the ground (Smathers and Gardner 1979; Smith 1985). Lack of ground cover may be due to dense shade produced by firetree canopies, or to some allelopathic effect of firetree leaves (Smith 1985); a relative of firetree native to Florida *(Myrica cerifera)* is known to produce allelopathic substances (Ewel et al. 1982). More insidious is the ability of firetree to fix nitrogen and invade nutrient-poor volcanic substrates of recent origin. On recent cinder substrate, firetree litter adds about four times more nitrogen to the soil than is derived from all other natural nitrogen sources combined (Vitousek et al. 1987).

The addition of large quantities of nitrogen alters the nutrient balance of entire ecosystems and thus may eventually encourage the invasion of nonnative plants that would not otherwise be able to grow as well as native plants adapted to naturally low nitrogen levels in developing Hawaiian soils (Vitousek, in press). Nitrogen-fixing firetree grows far more quickly than does the native 'ōhi'a, and firetree probably competes with 'ōhi'a for both water and sunlight (Smathers and Gardner 1979).

Firetree was formerly considered to be a noxious weed by the Hawaii State Department of Agriculture (1962) before populations became "uncontrollable" and is of little value as forage on rangelands (Hosaka and Thistle 1954). The fruit of firetree is dispersed by alien birds, particularly Japanese white-eyes *(Zosterops japonica)* (Vitousek et al. 1987) and house finches *(Carpodacus mexicanus)*; some native birds also take the fruit (LaRosa et al. 1985). Seeds are also dispersed by the feral pig *(Sus scrofa)*, which may consume large quantities of fruit during periods of heavy production (C.P. Stone, unpubl. data).

Firetree can be killed by several treatments with herbicides (Santos et al. 1989c). The State has for more than 25 years been attempting to control firetree at a major infestation in the Hāmākua District of Hawai'i Island, with some success (Whiteaker and Gardner 1985). Hawaii Volcanoes National Park managers have had success in controlling firetree in relatively intact systems, called Special Ecological Areas, over approximately 3,850 ha (9,500 a) (Cuddihy et al. 1988). However, the species is so widespread in the Park and elsewhere that biological control is considered the only hope for large-scale, long-term control. Previous attempts at biological control of firetree with several introduced insects were largely ineffectual (Clausen 1978). However, recent trips to the native habitat of firetree have resulted in the identification of potential insect and fungal control agents (Hodges and Gardner 1985; Markin et al., in press), and biological control research is ongoing. It is probable that the present range of firetree in Hawai'i, large as it is, is not the full extent of its potential range in the State (Whiteaker and Gardner 1985).

Approximately 15 additional alien tree species are listed by Smith (1985) as important invaders of native dry and mesic ecosystems. Some of these, such as New Zealand laurel *(Corynocarpus laevigatus)* on Kaua'i and *Leptospermum ericoides* on Lāna'i, are currently restricted to only one island. Other tree species are more widespread, such as silk oak *(Grevillea robusta)*, introduced from Australia about 1880 and planted in more than 40 forest reserves on five Islands (Skolmen 1979). Silk oak has become naturalized at least on Kaua'i, O'ahu, and Hawai'i (Wagner et al., in press), in dry regions between 350 and 1,600 m (1,150-5,250 ft) elevation (Smith 1985). Although silk oak is a potentially valuable timber species, it is particularly threatening to native plant communities because of its ability to form dense stands and to produce allelopathic substances, which prevent the establishment of other species (Smith 1985). The tree grows well on shallow, rocky soil in low rainfall areas (Nelson 1960); thus, it has been able to invade relatively recent substrates (e.g., in and near Hawaii Volcanoes National Park 'ōhi'a forests and shrublands).

Shrubs: Koa-haole, Lantana, Christmasberry, Gorse, Sourbush. One of the most widespread alien shrubs or small trees of the arid lowlands is *koa haole (Leucaena leucocephala)*, introduced from the Neotropics before 1837 and formerly cultivated for cattle feed and firewood (Wagner et al., in press). Seeds were broadcast over the lowlands, and now this species is found in dry habitats to above 700 m (2,300 ft) elevation on all the main Islands (Smith 1985). Koa haole was apparently kept somewhat in check by feral animals in the late 1800s, but with their reduction in many areas, the shrub came to dominate dry lowlands, forming dense stands as tall as 9 m (30 ft) (Egler 1942). In such monotypic stands all other plants are excluded (Smith 1985). Like several other successful invaders of dry habitats in Hawai'i, koa haole is a nitrogen fixer (Brewbaker and Styles 1984). Although often seen in very disturbed areas, the species also occurs in remnant dry and mesic forests that are important habitats for endangered and rare endemic plant species (e.g., on the upper slopes and ridges of the Wai'anae Mountains on O'ahu) (Wagner et al. 1985). Egler (1942) theorized that koa haole stands might be succeeded eventually by native dryland shrub and tree species, as had been observed to occur on Martinique, an island in the West Indies. Forty years later, Smith (1985) found no evidence for this in Hawai'i and considered such recolonization unlikely, because of the exhaustion of native plant seed banks. Recently, an introduced insect, the leucaena psyllid *Heteropsylla cubana*, has greatly reduced the vigor of koa haole stands (Smith 1985), but the economic value of the shrub precludes a biological control research effort. Koa haole has proven intractable to control efforts using herbicides, at least in Hawaii Volcanoes National Park (J.T. Tunison, pers. comm.).

Lantana *(Lantana camara)*, native to the West Indies and naturalized in many tropical countries, was introduced to Hawai'i in the mid-19th century and became naturalized before 1871 (Wagner et al., in press). Apparently, lantana spread rapidly after the introduction of alien fruit-eating birds such as the spotted dove *(Streptopelia chinensis)* and the common myna *(Acridotheres tristis)* (Clausen 1978). Today, lantana is distributed from sea level to above 1,070 m (3,510 ft) elevation, primarily in dry and mesic forests and shrublands (Wagner et al., in press) but also in wet habitats (Smith 1985). The shrub can form very heavy cover and produces allelopathic substances (Smith 1985), so it may be capable of displacing native shrub and herb species. Since lantana is toxic to cattle and has long been viewed as a pest of pastures and dry rangelands (Hosaka and Thistle 1954), it was an early target of biological control efforts. Just after the turn of the century, 23 insect species were introduced from Central America to combat its spread, and eight of these became established. Another round of biological control introductions took place in the 1950s and 1960s, with the result that the spread of lantana was halted in the drier regions of the Islands (Clausen 1978). Even though lantana cover was greatly reduced during the first half of the 20th century, it was apparently replaced, at least in arid areas, by other alien shrubs and grasses (Egler 1942).

Christmasberry *(Schinus terebinthifolius)*, native to Brazil, was introduced as an ornamental to Hawai'i before 1911 and now occupies dry and mesic habitats in the Islands from near sea level to approximately 920 m (3,020 ft) elevation (Wagner et al., in press). By 1962, Christmasberry had invaded 42,000 ha (103,740 a) in the State (Clausen 1978), and its lack of forage value and aggressive nature caused it to be considered a noxious pest in pastures (Hosaka and Thistle 1954). Christmasberry is capable of invading native dry and mesic forests and shrublands, and by forming dense stands it can shade out other plants; it also produces allelopathic substances (Smith 1985). As early as the 1940s, Christmasberry was recognized as an important invader of dry slopes on O'ahu (Egler 1942). Today, the shrub has replaced native forests and shrublands in much of the southern half of the Wai'anae Mountains, which have suffered disturbance from fire and grazing animals (Frierson 1973). In the Wai'anae Mountains, Christmasberry and other alien shrubs and vines threaten the last known population of the rare 'ōpuhe *(Urera kaalae)* (Obata 1986). Christmasberry is also a threat to lowland plant communities in Hawaii Volcanoes and Haleakala National Parks and Kaloko-

Honokohau and Kalaupapa National Historic Parks. Herbicidal control methods have been developed for Christmasberry, also a pest in U.S. mainland parks (Ewel et al. 1982). In Hawai'i, biological control insects have been introduced, but they have not been very successful against dense infestations (Clausen 1978).

Gorse *(Ulex europaeus)*, introduced early this century (before 1910) from western Europe, is now found primarily at high elevations on Maui and Hawai'i (Wagner et al., in press). In a recent survey of gorse distribution, the shrub was found over 8,260 ha (20,415 a) on the southeastern slopes of Mauna Kea, Hawai'i Island, to about 2,400 m (7,870 ft) elevation, with isolated pockets down to 450 m (1,480 ft). Gorse was also distributed over nearly 5,985 ha (14,789 a) on the northwestern slopes of Haleakalā, East Maui, between 630 and 2,220 m (2,070-7,280 ft) elevation. In portions of its range with heavy infestations, gorse forms dense, tall stands with as many as 60,000 stems/ha (24,290/a) (Markin et al. 1988). Although currently a problem on grazing lands, it is possible that the species may be able to invade open upland forests and subalpine shrublands. Of particular concern are the native high-elevation communities of Haleakala National Park, which are adjacent to gorse-infested rangelands (Loope et al., in press b). Gorse is a pasture pest in New Zealand, but there, when protected from fire and sheep browsing, the shrub can be succeeded by native forest trees (Allen 1936). Biological control research on gorse is under way, and two insects have been introduced to attack it: the gorse seed weevil *(Apion ulicis)* (Clausen 1978) and, more recently, a gorse moth *(Agonopterix ulicitella)* (Critchlow 1988).

Sourbush *(Pluchea symphytifolia*, formerly called *P. odorata)*, a tropical American species, was introduced to Hawai'i around 1931, probably accidentally (Hosaka and Thistle 1954). Within 20 years, it had become very common in the dry leeward low-lands of O'ahu (Fosberg 1948b), and by 1962, 20,000 ha (49,400 a) throughout the Islands were infested with sourbush (Clausen 1978). Currently, this fast-growing shrub occurs on the main Islands from sea level to above 1,000 m (3,280 ft) elevation (Smith 1985). While not generally considered an invader of intact ecosystems on a par with such disruptive species as Christmasberry or koa haole, sourbush can be locally devastating to coastal plant communities and has been blamed for hastening the aging process of anchialine pools (brackish pools with no surface connections to the ocean) through input of massive amounts of litter (Chai et al., in press).

Vines and Herbs: German Ivy, Nasturtium, Coccinea, Mullein. While vines and herbs are typically less important than other life forms as invaders of dry systems, one vine in particular has great disruptive potential. German ivy *(Senecio mikanioides)*, an ornamental from South Africa, was apparently naturalized in Hawai'i by 1910. Since then, it has become common in North and South Kona Districts and on lee-ward slopes of Mauna Kea, Hawai'i Island; the vine also occurs on Maui (Wagner et al., in press). In a survey of the island of Hawai'i 8 to 10 years ago, German ivy was found between 500 and 2,500 m (1,640-8,200 ft) elevation, primarily on the leeward slopes of Mauna Loa and Hualālai. Although sighted on less than 2% of the area sampled, Jacobi and Warshauer (in press) predicted that German ivy may be capable of estab-lishing on more than 63% of the lands covered by the U.S. Fish and Wildlife Service For-est Bird Survey, which would encompass vast acreages including most of Hualālai, both leeward and windward slopes of Mauna Loa, the summit area of the Kohala Mountains, and upper elevations of Mauna Kea. The impacts of this vine on native plants have not been well studied, but its habit of growing densely into the canopy of native trees (Smith 1985), such as *māmane (Sophora chrysophylla)* in upper-elevation forests, probably results in structural damage and reduction of available light. German ivy can form significant ground cover in native *koa/'ōhi'a (Acacia koa/Metrosideros polymorpha)* forests of South Kona on Hawai'i (Clarke et al. 1980), where it may interfere with native tree and shrub reproduction. Although mechanical control may be possible in localized areas (Smith 1985), if the predictions of Jacobi and Warshauer about *Senecio* distribution are realized, this vine will be a tremendous problem in native dry and mesic systems in Hawai'i in the future.

Another vine with disruptive potential in mesic forests is nasturtium *(Tropaeolum majus)*. An herbaceous ornamental native to South America, nasturtium has been naturalized in Hawai'i since the 1870s and now occurs at middle elevations on the islands of Kaua'i, Moloka'i, Maui, and Hawai'i (Wagner et al., in press). At present, this vine is a problem only in very localized areas, such as the montane mesic forest of Kīpuka Puaulu, Hawaii Volcanoes National Park. Prior to control efforts, nasturtium formed a monospecific cover in forest openings, grew into native tree canopies, and smothered small trees and shrubs (Tunison et al., in press).

A much more recent invader is the vine *Coccinea grandis* or scarlet-fruited gourd, a native of Africa, Asia, and Australia (Wagner et al., in press). Apparently introduced to the State in 1969 and collected first on O'ahu in 1985, *Coccinea* has become established at several O'ahu localities below 150 m (490 ft) elevation as well as near Kailua-Kona, Hawai'i Island, where it grows quickly, "smothering ground, shrubs, and trees in a solid blanket" (Linney 1986). While early naturalized populations were in disturbed lowland scrub, the vine has recently invaded native dryland vegetation containing endangered plant species (Linney 1989).

The high-elevation mountain slopes of Maui and Hawai'i have suffered less from alien plant invasions of the last 200 years than have the lowlands. However, during this century, common mullein *(Verbascum thapsus)* has become well established and abundant in Hawai'i's native-dominated subalpine ecosystems. Mullein is a native of Europe, widely naturalized in North America, and was first reported in Hawai'i in 1932, near the summit of Hualālai (Lyon 1932; Wagner et al., in press). During the next few decades, mullein became established on Mauna Kea and Mauna Loa and now occurs over an area of 2,000 km^2 (770 mi^2) (Juvik and Juvik, in press). Mullein has also recently (1986) invaded the slopes of Haleakala National Park, where two plants were found and removed from roadsides above 2,070 m (6,800 ft) elevation (Anon. 1989c; B. Gagné, pers. comm.). In a recent survey of roadsides on Mauna Kea, mullein was found between 1,625 and 3,300 m (5,330-10,820 ft) elevation, with densities of 160-190 plants per 100 m^2 near its lower limit (Juvik and Juvik, in press). These high densities may be related to the disturbed nature of the area sampled, but as mullein is reputedly distasteful to animals (Juvik and Juvik, in press), it may have had an unnatural advantage over native plants during past years when large numbers of feral sheep and goats roamed Mauna Kea. Juvik and Juvik (in press) speculated that mullein has usurped the niche of the endangered Mauna Kea silversword *(Argyroxiphium sandwicense* subsp. *sandwicense)* subsequent to that rare plant's reduction by feral animals. Mullein is likely to be a permanent component of the vegetation in heavily invaded areas; the seeds of mullein and other species of *Verbascum* can germinate after 100 years of burial (Kivilaan and Bandurski 1981).

Grasses: Broomsedge, Molasses Grass, Fountain Grass, Kikuyu Grass. Hundreds of grass species have been introduced into Hawai'i over the last 200 years (St. John 1973), and many have become established in dry rangelands and undeveloped leeward areas. Several grasses are notable for the amount of area they now cover and the great changes they have wrought in the plant communities they have invaded.

One of the most abundant alien grasses, at least on Oʻahu and Hawaiʻi, is broomsedge *(Andropogon virginicus)*. This tall bunchgrass, native to eastern North America, was first collected on Hawaiʻi in 1924 (Wagner et al., in press). By 1932, the grass was well established in the Kohala District of Hawaiʻi Island, where it spread from the Kohala Ditch Trail, the presumed site of introduction (Bryan 1977b). Broomsedge has become an important component of many lowland grasslands, ridge tops, and dry and mesic forests and shrublands. The grass ranges to above 1,600 m (5,250 ft) elevation, and at lower and middle elevations it is often the dominant ground cover (Smith 1985). The rapidity of broomsedge invasion may be illustrated by its increase over time in Hawaii Volcanoes National Park. In a 1947 checklist of exotic plants in the Park and a 1959 survey of the Kalapana extension, the grass was not even listed as present (Fagerlund 1947; Stone 1959); but a few years later, broomsedge was abundant from near sea level to 1,370 m (4,500 ft) elevation (Fosberg 1966). Currently, the grass is a dominant or co-dominant ground cover over thousands of hectares in the Park's coastal lowland and middle-elevation dry zones (Parman and Wampler 1977; J.T. Tunison, pers. comm. 1989). At some lowland sites, broomsedge apparently displaced the formerly dominant, native *pili* grass *(Heteropogon contortus)* (Stone 1959).

Perhaps the most important effects of this grass and the very similar bush beardgrass *(Schizachyrium condensatum,* syn. *A. glomeratus)* are the ability to carry fire, and adaptations to increase cover and range following fires (Sorenson 1977; Smith et al. 1980). In wet areas on Oʻahu, broomsedge has been found to increase erosion because it generally is dry and does not transpire during the heavy rainfall periods of winter (Mueller-Dombois 1973). Broomsedge is allelopathic to other plants (Rice in Smith 1985), is an efficient C_4 photosynthesizer (Rundel 1980), and is capable of growing well in nutrient-poor soils and under drought conditions (Sorenson 1977). Herbicidal control is not deemed practical for large stands of broomsedge, and biological control research is considered unlikely in the near future (Gardner and Davis 1982).

Molasses grass *(Melinis minutiflora)*, native to Africa, was intentionally introduced into Hawaiʻi in the early 1900s as cattle fodder (Wagner et al., in press). Subsequently, the grass spread into dry and mesic systems at low and middle elevations (Whitney et al. 1939). Besides being used as forage, molasses grass was planted out in dry areas of Molokaʻi and Lānaʻi for erosion control (Munro 1930). This grass produces dense, perennial mats that are capable of smothering other plants (Smith 1985) and preventing seedling growth and native tree reproduction (Scowcroft and Hobdy 1986). Perhaps the most serious impact of fire-adapted molasses grass is its ability to carry fire into areas with native woody plants. While much of the area invaded by molasses grass is in the disturbed lowlands, it also grows on the sides of gulches containing remnant stands of dry and mesic forest with candidate endangered plant species, at least on Molokaʻi (Cuddihy et al. 1982b) and Oʻahu (The Nature Conservancy of Hawaii, unpubl. data). Molasses grass has also invaded Haleakala National Park on Maui, where in the aftermath of goat control, it has "spread explosively" and developed heavy cover locally. Because of its tendency to spread and fuel intense fires, molasses grass presents a serious threat to rare dry forest plants and all other native vegetation in Kaupō Gap of Haleakalā (Loope et al., in press b).

Fountain grass *(Pennisetum setaceum)*, also native to Africa, is another fire-adapted bunchgrass that has dramatically spread since its introduction as an ornamental into Hawai'i in the early part of the 20th century. Fountain grass has invaded bare lava flows and open areas on Kaua'i, O'ahu, Lāna'i, and Hawai'i from sea level to an elevation above 2,000 m (6,560 ft) (Wagner et al., in press) and has been present on Maui for at least 25 years, the legacy of an ornamental planting in Wailuku (Loope et al., in press b). Fountain grass is already a dominant ground cover in dry ranchlands and open vegetation of North Kona and South Kohala on the island of Hawai'i. Based on environmental conditions in its present range, Jacobi and Warshauer (in press) predicted that fountain grass will be able to invade much more area on the upper-elevation slopes of Mauna Loa and Mauna Kea.

Fountain grass is particularly insidious because it is able to invade lava flows previously dominated by native plants, where it interferes with native plant regeneration, upsets natural succession, and allows damaging fires to occur where native vegetation alone would not have supported fires (Tunison et al. 1989). Fountain grass, by increasing the likelihood of fire, is recognized as a serious threat to endangered plant populations in dry forests on Hawai'i Island, for example *Kokia drynarioides*, *Caesalpinia kavaiense*, *Stenogyne angustifolia*, *Haplostachys haplostachya*, and *Lipochaeta venosa* (Wagner et al. 1985). Fears of fire in the habitat of endangered trees and other rare species (Powell and Warshauer 1985a) proved well founded when much of the proposed Pu'uwa'awa'a Natural Area Reserve, infested with fountain grass, burned in 1986.

Fountain grass has proven very difficult to manually control in Hawaii Volcanoes National Park, where it infests more than 8,000 ha (19,760 a) of the coastal lowlands. After more than 10 years of concentrated effort at the center of the Park's infestation, the species had not been eradicated, and Park managers developed a new strategy of confining the grass to the largely disturbed lowlands rather than attempting to remove it throughout its range in the Park (Tunison et al. 1989). If left unmanaged, it is feared that fountain grass could become distributed over all Park lands not covered by rain forest (Tunison, in press b).

On Maui, the State Department of Agriculture has successfully confined fountain grass to two small populations through persistent efforts in manual control, and the number of young plants recruited from the seed bank appears to be declining (Loope et al., in press b). If left untreated, fountain grass would be a serious threat to Haleakalā Crater and other upslope natural areas.

Kikuyu grass *(Pennisetum clandestinum)*, another African species, has been widely planted in upland pastures in Hawai'i (Hosaka 1958), where it may grow up to 3,050 m (10,000 ft) elevation (Loope et al., in press b). This aggressive grass has also invaded dry and mesic habitats as well as disturbed wet forests on all the Islands, where it forms thick mats, spreads rapidly by stolons, and produces allelopathic substances (Smith 1985). Of particular concern are rich assemblages of native species, such as the dry forest at Auwahi, East Maui, where kikuyu grass has formed a dense mat that prevents the reproduction of native tree species, some of them extremely rare (Medeiros et al. 1986). Kikuyu grass, along with other alien grasses, is also a problem in Kaupō Gap of Haleakala National Park, where it has been expanding upslope from ranches for more than 40 years (Loope et al., in press).

Even in the largely disturbed leeward lowlands, alien grasses may threaten remnants of native vegetation. In a recent study of one of only two known populations of the rare Hawaiian water fern *Marsilea villosa*, several grasses were found to share the fern's restricted habitat in a small crater on Koko Head, O'ahu. Of these alien species, ricegrass *(Echinochloa colona)* was recognized as the most serious competitor of *Marsilea* (Wester and Ikagawa 1988). At present, this site is managed by The Nature Conservancy of Hawaii, and volunteer groups are monitoring plant cover and removing encroaching alien grasses.

Fire

Natural Fire Regime and Fire History. Vogl (1969) proposed that naturally occurring fires, primarily from lightning strikes, have been important in the development of the original Hawaiian flora, and that many Hawaiian plants might be fire adapted. While accepting lightning as a potential ignition source, Mueller-Dombois (1981a) pointed out that most natural vegetation types of Hawai'i would not carry fire before the introduction of alien grasses. Native plant fuels typically have low flammability (Smith and Tunison, in press). However, fire probably influenced evolution of the montane ecosystems of Maui and Hawai'i, which contain grasslands of the native *Deschampsia nubigena* and stands of native shrub species and *koa (Acacia koa)*. As for rain forest vegetation, Mueller-Dombois (1981a) cited evidence from charcoal layers in soil pits that indicates an extremely low frequency of fire once every 700-1,000 years prior to human occupation of the Islands. Since carbon in soil layers may also represent lava flows or phreatic eruptions rather than free-burning fires, the true incidence of natural fire may be even rarer (Smith and Tunison, in press). Smith and Tunison concluded that although natural fire regimes in Hawai'i are "difficult to reconstruct," they are for most areas "best characterized as fire-independent."

In historic times, a few large rain forest fires have been reported, although it is usually unclear whether burned forests were intact or contained aliens that might have encouraged fires. During a drought in 1901, a fire burned an area 24 km (15 mi) long and 3-6 km (2-4 mi) wide in the southern part of the Hāmākua District of Hawai'i Island (Hall 1904). The composition of the forest was described as *'ōhi'a (Metrosideros polymorpha)* and koa with an understory of ferns, which after the fire was apparently followed by a "growth of weeds." This fire is said to have burned for "months" and destroyed forests over an estimated 12,150 ha (30,000 a) (Bryan 1961a). Other large fires apparently occurred in Hilo and Hāmākua forests in the middle of the 19th century, and the native forest trees were able to regenerate (Hall 1904, Horner 1908). Another long-burning fire occurred near Kula, Maui, in the 1880s and reportedly burned for weeks (Nelson 1967).

In the early decades of the 20th century, forest fires in Hawai'i were usually blamed on the flammability of the indigenous matted fern *uluhe (Dicranopteris linearis)*, which was considered a pest (Judd 1931) and thought to be a nonnative introduction (Bryan 1926b; Hostetler 1970). In 1926, an uluhe-fueled fire burned for a week, destroying much of the Pana'ewa Forest Reserve near Hilo and causing businessmen and homeowners of Hilo to fear for their property (Chamber of Commerce of Hilo 1926). This forest had somehow escaped development during the Hawaiian period despite its proximity to agricultural lands of Hilo (McEldowney 1979) and had been dominated, at least in its interior, by native trees when Isabella Bird passed through in 1873 (Bird 1966). After the fire, the burned area was planted with alien trees (Bryan 1926b), which today dominate most of the remaining forests of the Pana'ewa area.

Although large fires have certainly occurred in predominantly native vegetation (see Smith and Tunison, in press, for evidence of a large fire on Mauna Loa slopes 400 years ago), widespread invasion of alien fire-adapted grasses in the early 20th century (e.g., fountain grass *(Pennisetum setaceum)*, broomsedge *(Andropogon virginicus)*, molasses grass *(Melinis minutiflora))* is primarily responsible for the great increase in wildland fires in the second half of the century. Within a few decades of introduction, these grasses and others became well established in open forests and shrublands, as well as in the already-disturbed lowlands. Undoubtedly, an increasing human population and greater access to natural areas probably also played a role in increasing fire frequency.

Very few grasses were introduced in the first half of the 19th century (Nagata 1985), probably because large commercial ranching did not become well established until much later in the century. As ranches continued to be developed and expanded, many grass species were introduced and tried as forage plants (Hitchcock 1922). The period from 1906 to 1936 saw the introduction or first collection of most of Hawai'i's range grasses (Whitney et al. 1939). Even grasses of little forage value were brought in accidentally with the seed of palatable species. In some ranchlands, such as Kaupō on Maui, large areas were deliberately burned and seeded with pasture mixes (C.W. Smith, pers. comm. 1989).

Wildfires have increased in both size and number throughout the 20th century. Between 1904 and 1939, there were 205 forest and brush fires reported in the Territory; these burned approximately 14,780 ha (36,500 a), a yearly average of 420 ha (1,044 a). By contrast, in the period from 1940-76, at least 878 fires occurred, burning a total of 83,660 ha (206,650 a), or an average of 2,320 ha (5,740 a) per year, nearly a six-fold increase over the area burned in the previous 35 years (Schmitt 1977). In the 25-year interval between 1921 and 1945, foresters of Hawai'i Island recorded only 61 forest fires (Bryan 1947). Forty years later (1986), a total of 73 fires occurred in just one year on Hawai'i Island (Hawaii State Department of Business and Economic Development 1988), and the total area of forest and shrubland burned in the State exceeded 10,100 ha (25,000 a).

This same pattern of increasing fire size and frequency has been observed in Hawaii Volcanoes National Park, where accurate records of fires have been kept since 1924. The Park supports large stands of relatively dry 'ōhi'a forest and native shrublands in a lowland zone between 305 and 1,070 m (1,000-3,500 ft) elevation; below this is a coastal area previously modified by Hawaiians. The ground cover in both these lowland zones is now dominated by alien grasses. Before the establishment of alien grasses, particularly broomsedge, bush beardgrass *(Schizachyrium condensatum)*, and molasses grass *(Melinis minutiflora)*, fires were small and infrequent, with only 27 fires averaging 4 ha (10 a) in size recorded from 1924 to 1967 (National Park Service 1989; J.T. Tunison and J. Leialoha, unpubl. data). By the 1960s, broomsedge and other grasses had invaded the Park (Fosberg 1966), but a large population of feral goats *(Capra hircus)* kept grasses cropped and maintained the dominance of annual grasses over tall, perennial species (Mueller-Dombois 1981b), at least in the central and western coastal lowlands. When goats were removed from most of the Park in the 1970s, tall perennial grasses such as broomsedge and bush beardgrass came to form a continuous ground cover in formerly open vegetation in that part of the coastal lowlands with sufficient soil. There was a great "increase in biomass and a fuel bed capable of supporting fire" (National Park Service 1989), and the sizes of fires increased two orders of magnitude over those recorded prior to 1970 (Smith 1985). More than 90% of all fires that have occurred in the Park's coastal lowlands took place after the feral goat removal program began (National Park Service 1989). A similar problem now exists in western Kaupō Gap of Haleakala National Park on Maui, where the increase in biomass of alien grasses following feral goat removal has resulted in greatly increased potential for wildfire, threatening the ongoing recovery of native trees and shrubs (A.C. Medeiros and L.L. Loope, unpubl. data).

Grasses, particularly those that produce mats of dry material or retain a mass of standing dead leaves, provide fine fuels that allow fires to carry into native forests and shrublands that they have invaded; this vegetation would not otherwise easily burn. Broomsedge is known to be extremely flammable with a very high ratio of surface area to volume, and molasses grass can produce extremely combustible mats greater than 2 m (7 ft) deep (Fujioka and Fujii 1980). During fires observed in Hawaii Volcanoes National Park, broomsedge has been found to burn at very high humidities and fuel moistures (National Park Service 1986a).

In the two decades between 1968 and 1988, 58 fires were recorded in Hawaii Volcanoes National Park, and the average size of fires had increased 50-fold over the previous four decades to 204 ha (500 a). More than half of these fires occurred in the lowland and montane dry zones of the Park, between 305 and 1,100 m elevation (1,000-3,600 ft), an area heavily invaded by broomsedge and bush beardgrass (National Park Service 1989). Most recent fires have been caused by human carelessness, but several have been ignited by lightning. The best-documented lightning-ignited fire burned approximately 4,250 ha (10,500 a) in 1986 before being extinguished by Park personnel (Tunison and Leialoha 1988). Lava flows are, of course, another natural ignition source and were the cause of 26% of Park fires from 1924 to 1988 (National Park Service 1989). In upland wet forests, fires rarely carry long distances from lava flow edges (Vogl 1969), and wildfire becomes likely only when flows reach the lowlands infested by alien grasses.

Fire Effects. The effects of fires on native Hawaiian vegetation are largely deleterious. Native woody plants may recover from fire to some degree, but fire tips the competitive balance toward alien species (National Park Service 1989). Although a few native plants, such as *koa (Acacia koa)*, respond favorably to fire by resprouting from suckers, or by producing prolific seedlings, as does *'a'ali'i (Dodonaea viscosa)*, they do not require fire to survive, and such adaptations may relate to other stress factors or simply be a trait inherited from ancestral stock (Mueller-Dombois 1981a). In numerous studies carried out in Hawaii Volcanoes National Park and elsewhere, fire has generally been shown to adversely affect native woody plants (National Park Service 1989; Smith and Tunison, in press). *'Ōhi'a (Metrosideros polymorpha)* is the most abundant tree in dry lowland and mid-elevation regions that still support native woody vegetation. Such areas are often heavily invaded by alien grasses and are prone to fires. In several studies of burns in 'ōhi'a woodland, only about 50% of the trees regenerated by resprouting (Parman and Wampler 1977; Tunison et al., in press). Fire intensity is an important factor in the survival of 'ōhi'a; tolerance and survivability decrease as intensity increases (Smith and Tunison, in press). Small 'ōhi'a are more likely to be killed by fire than are large ones with multiple trunks (Smith et al. 1980), but repeated fires in an area may eventually result in the total loss of 'ōhi'a and inhibit reestablishment of most native woody species (Mueller-Dombois 1981a). Several native tree species of lowland forests, such as sandalwood *(Santalum* sp.), *kōlea (Myrsine lessertiana)*, and *alahe'e (Canthium odoratum)*, may survive fires if they are not too severe (Warshauer 1974).

Mature rain forests do not seem to be particularly vulnerable to fire, and some common understory plants such as tree ferns or *hāpuʻu (Cibotium* spp.*)* are known to have a high fire survival rate. However, in rain forests that have burned, disturbance by fire is usually followed by invasion of alien shrubs and grasses (Smith and Tunison, in press).

Several common native shrubs of the lowlands are capable of surviving low-intensity fires and resprouting from the base, particularly *ʻūlei (Osteomeles anthyllidifolia)*, *ʻaʻaliʻi*, and *ʻōhelo (Vaccinium reticulatum)* (Warshauer 1974), although regrowth may be slow (Smith and Tunison, in press). Because ʻaʻaliʻi produces many seedlings following fire, its density may actually increase in burned areas (Hughes 1989; Tunison et al., in press). Other native shrubs, particularly *pūkiawe (Styphelia tameiameiae)*, are very susceptible to fire. Burned low-elevation shrublands containing pūkiawe typically show a great decrease in cover or even a total loss of the species, and pūkiawe seedlings are rarely observed in fresh burns (Warshauer 1974; Parman and Wampler 1977). In mid-elevation shrublands of Kaupō, Maui, fire threatens to retard recovery and succession of native woody species following feral goat *(Capra hircus)* removal (A.C. Medeiros and L.L. Loope, unpubl. data).

Unlike native woody species, many alien grasses recover quickly and increase in cover following fires. This pattern is seen in most fire studies in Hawaii Volcanoes National Park, where fire-adapted grasses such as broomsedge, bush beardgrass, molasses grass, and thatching grass *(Andropogon virginicus, Schizachyrium condensatum, Melinis minutiflora, Hyparrhenia rufa)* invariably increase in cover following fires (Parman and Wampler 1977; J.T. Tunison and P.K. Higashino, unpubl. data; L.W. Cuddihy, unpubl. data). Hughes (1989) noted a pattern of increased cover of alien grasses and decreased cover of native shrub species that persisted after 18 years in a burned area near Kīpuka Nēnē campground.

Fire-adapted grasses such as broomsedge are able to recover quickly because fire stimulates growth from the base of clumps and encourages seed germination (Sorenson 1977). Broomsedge and other alien species such as molasses grass and fountain grass *(Pennisetum setaceum)* use the C_4 photosynthetic pathway, which allows them to efficiently photosynthesize at a high rate even in hot, dry habitats with low soil nitrogen levels (Sorenson 1977; Hughes 1989; Vitousek, in press).

Urbanization and Development

The loss of original vegetation cover due to the development of human habitations began with the Polynesians, who lived primarily in the coastal areas and lowlands. Far more significant than the development of their living areas and villages were the losses of forest resulting from conversion to agricultural fields and extensive burning, which were part of Hawaiian agricultural practices (Kirch 1982).

The human population of the Hawaiian Islands when first contacted by Europeans in 1778 was probably between 200,000 and 1 million; varying estimates were made in the 18th century (Schmitt 1977; Stannard 1989). This population greatly decreased in the decades following contact; despite the influx of people from Europe and North America, the human population of the Islands did not reach the quarter-million mark again until 1920 (Schmitt 1977), although it had begun to rise as early as 1870 (Fisher 1973). As late as 1853, the population of Hawaiʻi was still primarily restricted to the coastal lowlands near fishing and farming sites, and the interiors of the Islands were almost uninhabited (Coulter 1931). In 1820, the towns of Hilo, Kailua, Lahaina, and Hāna each had populations of about 2,000, and Honolulu was only slightly larger; 33 years later, Honolulu had more than 11,000 people, which included most of the Kingdom's foreigners (Chow 1983). In the 20th century, cities such as Honolulu and Hilo began to grow more rapidly, but it was not until 1930 that the numbers of people living in urbanized areas exceeded the rural population (Schmitt 1977). McEldowney (1979) attributed the increase in the population of Hilo, the second-most populous city in the Hawaiian Islands, to the relocation of sugar plantation labor after nullification of labor contracts during the annexation of the Hawaiian Kingdom. The growth of Honolulu was also influenced by the migration of ex-plantation laborers (Chow 1983).

The primary focus of 20th century urbanization in the Islands has been Oʻahu, which in 1980 had 41,960 ha (103,640 a) of urban land, more than half of the State's total (75,160 ha or 185,640 a) (Armstrong 1983). Honolulu has been the population center of the Hawaiian Islands since the late 1800s (Schmitt 1977), when the population of Oʻahu finally surpassed that of Hawaiʻi Island (Chow 1983). In 1980, the city contained nearly 40% of the people in the State (Armstrong 1983). Although most of the valleys and the coastal plain of Honolulu had already been disturbed by activities of pre-European contact Hawaiians, 20th century urbanization and population growth have expanded development to many areas not used by Hawaiians for agriculture, such as ridges, slopes, and the upper reaches of valleys. Several residential subdivisions were developed on Honolulu ridges and slopes by 1920 (e.g., St. Louis Heights, Diamond Head) (Chow 1983). Certainly the development of a modern city has transformed the landscape to a far greater extent than the Hawaiians were capable of doing before European contact. Urbanization of greater Honolulu has not only replaced the land's former vegetative cover with structures, roads, and alien plantings (Wester 1983) but has even obliterated such natural features as Salt Lake (Āliapaʻakai), formerly the largest natural lake in the Hawaiian Islands (Cooper and Daws 1985). By 1969, more than 3,240 ha (8,000 a) of land on Oʻahu had been converted to streets and highways (Schmitt 1977). Urbanization results in the loss of nearly everything that lived in the area before development. Gagné (1981) pointed out that several native insect species, which were historically collected in areas that became part of the city or suburbs of Honolulu, are undoubtedly extinct today.

For several decades, the fastest growth rates on Oʻahu have been in suburban areas on the edge of the central city. This expansion (or urban sprawl) began after World War II and involved such areas as Pearl City and the windward towns of Kailua and Kāneʻohe (Chow 1983). Because of this great increase on windward Oʻahu, a tunnel through the Koʻolau Range to connect these communities with Honolulu was necessary by the 1950s, and later a second tunnel was constructed. By the 1970s, a third tunnel and "H-3" freeway were proposed. Construction of the H-3 was opposed by environmentalists who were concerned about the loss of native plants, particularly localized species of *Cyrtandra* and *loulu* palms *(Pritchardia)*, in Moanalua Valley, the original route proposed for the Highway (Gagné 1974). More than 10 years later, the proposed route for H-3 had been changed to a corridor from Haʻiku Valley on windward Oʻahu through the Koʻolau Mountains to Hālawa Valley on the leeward side of the island. This route also met with resistance from conservationists and biologists, because it would result in the destruction of habitat for the endangered Oʻahu creeper *(Paroreomyza maculata)* and endangered land snails *(Achatinella* spp.) on Conservation-zoned land in the Koʻolau Mountains. The highway would also destroy vegetation and the stream of the valley floor in Hālawa (Gagné and Montgomery 1986). While H-3, "probably the nation's longest-running highway dispute" (Myers 1976), has been the most

95

controversial highway project in recent years, it is unlikely to be the last proposed highway or road to severely impact native ecosystems. Pressures to develop and improve transportation are inevitable as population growth and urbanization continue.

Tourism and Resort Development. Tourism, today the State's most important industry, began in Hawai'i in the middle of the 19th century with the first hotel built on the island of Hawai'i near Kīlauea Crater in 1866. By the turn of the century, the focus of tourism and hotel-building had moved to Waikīkī on O'ahu (Morgan 1989). The number of visitors to Hawai'i grew slowly during the first half of the 20th century, from less than 10,000 in 1922 to more than 100,000 in 1955 (Schmitt 1977). Tourism boomed after World War II, helping to prevent a predicted economic decline (Baker 1961). Statehood provided another tremendous boost to the industry, as did the development of jet travel and the reduction in airline fares (Morgan 1989). Annual visitor numbers began to rise dramatically just before Statehood in 1959, doubling about every five years throughout the 1960s and early 1970s (Schmitt 1977). Within 14 years of Statehood, tourism increased five-fold, and the value of constructed hotels grew 18-fold (Fisher 1973). In 1987, nearly 6 million tourists visited the State; at any given time during the previous year, visitors represented approximately 15% of the State's population (Morgan 1989). To accommodate this increasing influx of visitors in the last three decades, there has been a boom in hotel building, mostly on O'ahu and Maui, the two islands that have seen the most urban development of all the Islands (Cooper and Daws 1985). By 1982, almost 58,000 hotel units had been built (Armstrong 1983), and by 1988 this had increased to more than 69,000 (Hawaii State Department of Business and Economic Development 1988).

While many hotels are built in areas already urbanized, new resorts are primarily located on the coast, and their effects reach far beyond the buildings constructed to house tourists. The building of hotels and the creation of resorts in an area leads to increased urbanization, commercial development, and residential subdivision building. This is evident in the former fishing village of Kailua-Kona on the island of Hawai'i, where tourism and real estate are now the primary commercial activities. Urbanization has increased rapidly there; one indication is the doubling of the number of real estate agents in Kailua (from 200 to 400) in just three years (Kelly 1983). West Hawai'i has been the fastest growing region on the Island (Culliney 1988), and resort development has also accelerated growth on Maui, Moloka'i, and Kaua'i (Chow 1983). The State has used investments to encourage tourist development, particularly on the main Islands other than O'ahu (Myers 1976). Recently, "huge, upscale resorts" have been built on Lāna'i and Moloka'i, where little tourism occurred in the past (Culliney 1988). A harbinger of even greater change on Lāna'i was the 1970s reclassification of 6,880 ha (17,000 a) of Conservation land to Urban, earmarked for tourist development (Myers 1976).

A recent trend is the creation of destination resorts far from urban areas and other hotels. The first of these was developed in 1961 at Kā'anapali, near Lahaina on Maui (Cooper and Daws 1985). Some resorts, particularly on islands other than O'ahu, are constructed in dry, sunny areas that never supported large populations in the past. Construction at remote sites results in increased access to an area and may cause destruction of natural areas miles away from resorts, including unseen damage to marine systems (Culliney 1988). Important remaining lowland systems may be sacrificed to resort development; a recent example is the loss of one-third of the known anchialine pools in the State, when an elaborate resort complex was built at 'Anaeho'omalu, North Kona, on Hawai'i Island (Culliney 1988). (Anchialine pools are unusual near-shore tidal pools with no direct surface connection to the sea. They contain a high percentage of endemic organisms and are often important sites for native waterbirds.) Although a legal appeal was being heard in Circuit court, the developer bulldozed 70% of the 'Anaeho'omalu ponds, destroying habitat for more than 60 species of plants and animals (Conant 1986). The 'Anaeho'omalu site was considered to be of exceptional natural value because of the number and diversity of its pools, which harbored a rich assemblage of invertebrates and fishes, as well as a rare species of eel *(Gymnothorax hilonis)* (Maciolek and Brock 1974). Another North Kona site with anchialine pools and endangered shore bird habitat was recently spared development when the landowner withdrew a request for rezoning (Mull 1987). Even though development may have been forestalled or delayed at this site, ten more resorts have been planned for the West Hawai'i coast (Mull 1986c).

A massive resort project has also been proposed for the rural Ka'ū District of Hawai'i Island. The project would convert a remote 1,310-ha (3,240-a) parcel currently zoned Conservation and Agriculture into an urban development that would quadruple the existing population of the District (Harada-Stone 1989). Drawing parallels with such densely populated and developed tourist destinations as Bermuda, Culliney (1988) warned that rapid and unbridled tourist development in Hawai'i may lead to great intensification of human population densities in the Islands and "acute environmental woes".

Hawai'i has had a comprehensive land use law with state-wide zoning and "permanent" boundaries since 1961 and was the first state to adopt such a planning tool (Myers 1976). In spite of this, developers have often found it possible to obtain rezoning to reclassify Conservation lands to urban or resort uses. Rezoning has been particularly pronounced on Hawai'i, which in the 1970s led all the other Islands in the amount of land rezoned to Urban use (Chow 1983). Mull (1987) warned of the negative, long-term consequences to native plants, wildlife, and communities when "great land tracts are transferred to urban use."

Even when rezoning is not accomplished, much development can take place on "Conservation" lands through the Conservation District Use application process. Such uses have included tourist attractions on O'ahu and an airport on Hawai'i; in a five-year period (1971-76) no public development on Conservation land was turned down because of an adverse Environmental Impact Statement (Myers 1976).

Upland Subdivisions and Other Development. Although the expansion of urban areas and resorts has taken place primarily in the lowlands, often in localities affected by pre-European Hawaiian land uses, the 20th-century development of residential and "agricultural" subdivisions has expanded such disturbance to many upland sites still covered by native vegetation. In just eight years (1960-68), the amount of land in the state classified as Urban doubled, primarily because of undeveloped subdivisions (Schmitt 1977). The focus of the most dramatic subdivision of undeveloped land has been the island of Hawai'i, where between the late 1950s and 1975 approximately 80,000 residential lots were created, primarily for speculation (Cooper and Daws 1985). While some of these subdivisions were developed on former agricultural lands or relatively recent lava flows with little vegetation, others, particularly in the Puna District, were in wet and mesic forests dominated by 'ōhi'a *(Metrosideros polymorpha)* or other native trees. One of the most flagrant cases of development for purely speculative purposes is that of Royal Gardens, a 730-ha (1,800-a) parcel downslope of the East Rift of Kīlauea and east of Hawaii Volcanoes National Park, which has, since 1983, been frequently featured in the news. Royal Gardens was subdivided first in the 1960s and was heavily promoted on the U.S. Mainland (Cooper and Daws 1985), despite its position in a volcanic hazard zone (Anonymous 1976). This subdivision is directly adjacent to a section of the Park notable for its rich mesic forests, uncommon plant species (Warshauer 1977), and unusually high lowland populations of native birds (Conant 1980). By 1985, much of the subdivision was covered by lava flows from the ongoing Kīlauea eruption, which began in 1983.

Many other Puna subdivisions were developed in native forests with a high diversity of native plant species, but the pre-disturbance cover can often only be gauged by what remains nearby. A survey of Malama-Kī Forest Reserve, adjacent to the Leilani Estates subdivision, showed that a candidate endangered tree species and a recently discovered *Cyrtandra* in closed *lama (Diospyros sandwicensis)* forest were present in an area that was previously continuous with the forest of the subdivided area (Clarke et al. 1981). A number of large parcels were subdivided between the villages of Mountain View and Volcano, bordering the large tract that later became the Kahaualeʻa Natural Area Reserve. Large subdivisions have also been developed in formerly forested lands of the districts of Kaʻū and South Kona; examples are a parcel contiguous to Manukā Natural Area Reserve (The Nature Conservancy of Hawaii 1987) and an area in the upland forests of Kapuʻa, which J.F. Rock considered one of the island's richest botanical sites (Rock 1913).

Most of these subdivided lots on Hawaiʻi Island have never been cleared or occupied; for example, only 5% of those in Puna had residences in 1983 (Cooper and Daws 1985). Even so, the process of bulldozing roads fragments and degrades existing vegetation and provides avenues for the establishment of alien plants. On Maui, weed introduction into the uplands has been linked to road construction and resurfacing (Loope et al., in press).

Twentieth-century development of upland forests has not been restricted to subdivisions and small agricultural parcels. In the 1940s, foresters unsuccessfully opposed the development of Kulani Prison Honor Camp in what was part of ʻOlaʻa Forest Reserve. Bryan (1944a, 1944b) decried the road-building and clearing in native wet forest because of the likelihood of weed introductions and the potential for fire. Subsequent recognition of the Kūlani area forests as important habitat for several endangered bird species (Scott et al. 1986) and the establishment of adjacent lands as a State Natural Area Reserve serve to underscore the inappropriateness of a large-scale institutional and agricultural compound at this upland site. Between Kūlani Cone and Volcano Village, large-scale upland agricultural development continued into the 1950s, when hundreds of hectares were removed from forest reserve status and sold as farm lots (Mull 1987). Today this incongruous parcel of farm and grazing land is surrounded on three sides by Hawaii Volcanoes National Park and State Natural Area Reserve and serves as a potential source of invasive alien plants.

The various branches of the U.S. military own or control a total area of nearly 98,000 ha (241,920 a) in the Hawaiian Islands, most of it on the islands of Oʻahu and Hawaiʻi (Armstrong 1983). Several Oʻahu military reservations cover much of the northern Koʻolau and Waiʻanae Mountains, areas that contain sites supporting rare natural communities and plants; some installations are located near existing or proposed State Natural Area Reserves (The Nature Conservancy of Hawaii 1987). The primary military site on the island of Hawaiʻi is Pōhakuloa Training Area, located near 1,830 m (6,000 ft) elevation in the saddle formed by Mauna Kea, Mauna Loa, and Hualālai. Pōhakuloa is notable as habitat for three endangered plant species *(Haplostachys haplostachya, Lipochaeta venosa, Stenogyne angustifolia)* and also supports populations of several candidate endangered plant species and important examples of montane dry forests (Higashino et al. 1977). Fire is one of the greatest threats to rare plants and native vegetation in military training areas such as Pōhakuloa (Wagner et al. 1985), where firing of weapons and large numbers of troops provide potential ignition sources. Other negative effects of military use are the accidental introduction of alien plant species with equipment or personnel (Smith 1989b) and the bulldozing of military roads, which may be useful as firebreaks but also act as corridors for the movement of weeds into stands of native vegetation. On Maui, the military have been implicated as a "significant source of plant introductions" at Haleakala National Park, where military use during World War II included the transport of materials from O'ahu (Loope et al., in press b). Perhaps the most alarming use the military has made of native forests occurred on Kaua'i, where testing of chemical defoliants, including Agent Orange, occurred in the 1960s (Culliney 1988).

Geothermal and Hydroelectric Development. Although Hawai'i has little heavy industry, and most manufacturing occurs on already heavily urbanized O'ahu (Armstrong 1983), the development of geothermal power threatens to bring industrial complexes to Agricultural and Conservation District areas of the Islands, at least on Maui and Hawai'i. Culliney (1988) predicted that "geothermal power production could become the key factor leading to virtually unlimited development." Exploratory geothermal wells were drilled on the active East Rift of Kīlauea on Hawai'i Island as early as 1961, and a successful well (HGP-A) was drilled there in 1975 and is now generating approximately 3 megawatts of energy (Thomas 1985).

In the early 1980s, a large geothermal development was proposed for Kahauale'a, a privately owned parcel on the upper East Rift of Kīlauea, directly adjacent to Hawaii Volcanoes National Park. Most of the more than 10,000 ha (25,000 acres) of land proposed for this industrial development was zoned Conservation and was covered by relatively intact 'ōhi'a *(Metrosideros polymorpha)* rain forest. The 250-megawatt project was envisioned as a complex of five power plants and up to 70 wells connected by a network of many miles of roadway and piping (Towill 1982). It was estimated that construction would require the clearing of 170 ha (420 a) of forest (Clark 1982). A general estimate of land required for structures and graded clearings in a geothermal complex is 0.1 to 1.2 ha (0.3-3 a) per megawatt (Hawaii State Department of Planning and Economic Development 1982).

In the Environmental Impact Statement prepared for the Kahauale'a project, the loss of land required for the complex was presented as insignificant to plants and wildlife, although it was recognized that development would impact the only known population of a rare fern *(Adenophorus periens)*, a candidate for Federal Endangered Species status. Several species of endangered birds, particularly the 'ō'ū *(Psittirostra psittacea)* and Hawaiian hawk or 'io *(Buteo solitarius)*, were also at risk (Towill 1982). Although an estimated 5-10% of *Adenophorus periens* plants might have been destroyed by construction and clearing, the vigorous population was thought to be capable of surviving such a loss (Clark 1982).

The development of an industrial complex in the upland native rain forest of Kahauale'a was vigorously opposed by conservationists and community associations, who cited the danger of "habitat degradation through an accelerated rate of exotic plant influx" and pointed out the lack of knowledge about the long-term effects of geothermal emissions on native plants and animals (Stemmermann 1983). The developers were

99

eventually given permission by the Board of Land and Natural Resources to drill exploratory wells. Subsequently, the 1983 Pu'u 'O'o eruptions covered many of the proposed well and power plant sites before drilling began, some to a depth of 18 m (60 ft). Eventually, the State exchanged Natural Area Reserve and Forest Reserve lands of the middle East Rift for Kahauale'a, and the Governor declared Kahauale'a a Natural Area Reserve (Culliney 1988). While the upper East Rift forests now appear secure from industrialization, the focus of geothermal development has simply moved downrift (Mull 1986d). Forests of Kīlauea's middle East Rift may be typically less dense and more disturbed than those at higher elevations, but stands of tall, closed-canopy, highly diverse rain forest containing populations of several rare plant species are found there (Char and Lamoureux 1985).

The East Rift of Kīlauea is not the only potential site for geothermal development in Hawai'i. One assessment identified 20 potential geothermal resource areas on five Islands (Thomas 1985), but only sites with good probability of a high-temperature resource are likely to be developed for electrical energy in the near future. In addition to the three sections of the East Rift (upper, middle, and lower), Hawai'i Island has potentially exploitable geothermal resources on the Southwest Rift of Kīlauea, the Northwest Rift Zone of Hualālai, and the Southwest and Northeast Rift Zones of Mauna Loa. On Maui, the sites with greatest geothermal potential are the East and Southwest Rift Zones of Haleakalā. Except for the Southwest Rift of Kīlauea, all these potential geothermal development areas contain Conservation-zoned land. Several of them support significant habitat for endangered bird species and rare plants, particularly the East Rift Zone of Haleakalā (which stretches between Hāna and Haleakala National Park) and the Northeast Rift of Mauna Loa, which includes the area around Kūlani Cone (Hawaii State Department of Planning and Economic Development 1986). Since 1983, it has been possible to designate geothermal resource subzones in any land use District, including land zoned Conservation (Hawaii State Department of Planning and Economic Development 1986).

In 1987, electric utilities on Hawai'i Island supplied an average of 66 megawatts of power hourly (Hawaii State Department of Business and Economic Development 1988), and the Island's peak power needs have reached 126 megawatts (Bishop 1989). Despite the relatively modest needs of Hawai'i Island, the development of geothermal complexes to produce 500 megawatts is envisioned in the near future if an interisland deep-sea cable to transport energy to O'ahu becomes feasible (Hawaii State Department of Planning and Economic Development 1982). The impacts on existing natural vegetation and native animal habitat, as well as the loss of agricultural land, will increase with the increasing energy-producing capacity of geothermal developments. The losses could be even greater if large-scale geothermal projects stimulate development of local industries to use the geothermal steam directly (Lew 1981) or provide impetus for the development of energy-consumptive, potentially polluting industries such as manganese nodule processing. A single refinery processing manganese ores into cobalt and nickel would cover hundreds of hectares with tailings and waste treatment ponds; proposed sites for refineries include the Puna and Kohala Districts of Hawai'i Island (Culliney 1988).

Hydroelectric power developments are another potential disturbance in native forests, although their impacts are perhaps less far-reaching than those of geothermal projects. Since 1964, hydropower has provided the state with 78-115 million kwh each year (Hawaii State Department of Business and Economic Development 1988). A few small hydroelectric plants have been in place on Kaua'i for 50 years, and recently much larger projects have been proposed for valleys on Kaua'i, Maui, and Moloka'i (Culliney

1988). Although at present only 2% of the electricity consumed on Hawai'i Island comes from hydroelectric power (Towill 1982), two new hydroelectric projects are planned for streams near Hilo: a 10-megawatt plant on the Wailuku River and a 14.6 megawatt development on nearby Honoli'i stream (Bishop 1989). While the Honoli'i project has generated more controversy for its possible adverse effects on a nearby surfing beach, the Wailuku River plant would impact 47 acres of State-owned Conservation District land (Bishop 1989) in an area notable as habitat for a rare variety of *'ōhi'a* (*Metrosideros polymorpha* var. *newellii*) and several endangered birds (Mull 1986c).

SUMMARY OF VEGETATION ALTERATION IN THE HAWAIIAN ISLANDS

Pre-European Contact Changes

By the time of Captain James Cook's arrival in the Hawaiian Islands in 1778, the original vegetation of the lowlands had been greatly altered by more than 1,000 years of Hawaiian occupation. Agricultural practices of the Hawaiians were the major cause of environmental change in the Islands. Kirch (1982) expressed the opinion that any lowland area receiving 500 mm (20 in.) or more of annual rainfall would reveal archaeological evidence of Hawaiian agricultural use if examined. In nearly every valley with a permanent stream, the natural vegetation was replaced with irrigated taro *(Colocasia esculenta)* and other plants introduced by Hawaiians. The development of large dryland field systems on the island of Hawai'i (and East Maui) completely cleared the original dry and mesic vegetation over large tracts of the lower leeward slopes. Newman (1972) estimated that as much as 260 km^2 (100 mi^2) of the island of Hawai'i (amounting to 2% of the land area) was "churned up" during preparation of field systems.

Shifting cultivation, using slash and burn techniques, was carried out at least to 460 m (1,500 ft) elevation on windward slopes and in areas near sites of irrigation agriculture. Plants introduced by the Hawaiians largely replaced native plants in succession on fallow and abandoned fields. Dryland cultivation greatly altered the vegetation cover of the islands of Ni'ihau, Kaho'olawe, and Lāna'i long before the arrival of Europeans.

Even in lowland areas not actually cultivated, natural vegetation was degraded through the use of fire to encourage thatching grasses (Kirch 1982), firewood gathering, and removal of timber for construction purposes. The forest zone above cultivated areas was used for wild plant products, canoe logs, and for collecting feathers and was invaded to some extent by plants of Polynesian introduction. Only above 760 m (2,500 ft) elevation was the original vegetation essentially untouched by Hawaiians (Kirch 1982). Lowland areas below this upper limit of Polynesian influence where native vegetation has survived to the present and escaped modification are found on rough substrates, steep terrain, or remote coasts of difficult access.

In recent years, evidence for the destruction of lowland vegetation has come from analysis of fossil bird bones and land snails. Olson and James (1982) identified 40 previously unknown extinct species of birds from fossil bone deposits at a number of sites (primarily lowland) on five Hawaiian Islands; this doubled the number of endemic land bird species known from Hawai'i. Additional work on fossilized bird remains from a cave on Maui tripled the number of bird species known to have occurred on that Island (James et al. 1987). Radiocarbon dates on the Maui fossils and both radiocarbon dates and stratigraphic analysis at the previously discovered sites indicated that many of these birds had become extinct since Polynesian occupation of the Islands. Olson and James (1982) also pointed out that the currently understood fossil record is probably incomplete and the "original bird diversity" of the Islands may have been even greater. On O'ahu, deposits of endemic land snails indicate changes in the snail fauna during the Polynesian period, with some endemic taxa decreasing and others, more adapted to disturbance, increasing (Kirch 1983). Both Olson and James (1982) and Kirch (1983) suggested that the primary reason for the extinction of native animals and change in faunal composition was the destruction of lowland forest habitats, particularly those of the dry leeward regions. (Predation by humans and introduced rats also had a role in the demise of native birds, especially flightless and ground-nesting species.)

Although the plant communities in much of the Hawaiian lowlands were disturbed, it is clear, from early botanical collections and the many native plants that were regularly used in the Hawaiian culture, that some plants rare or non-existent in the lowlands today were relatively common in Hawai'i before European contact. St. John (1976) found more than a dozen species that are now extinct among the plants collected in Kona by David Nelson in 1779. The disruption of montane vegetation and the almost complete loss of native plant species in the coastal and lowland zones that characterize present-day Hawai'i were only accomplished through modern land-use practices and the disturbances caused by alien plants and animals in the last 200 years.

Post-European Contact Changes

The alteration of natural vegetation in the Hawaiian Islands has progressed steadily upslope since 1778. Perhaps the most serious and persistent early impact of western culture was the introduction of large grazing and browsing mammals, which quickly established large feral populations and began to open up and destroy upland forests and high-elevation systems. Also significant in the early post-European contact period, particularly on O'ahu and near lowland population centers, was the wholesale exploitation of lowland forests for sandalwood (*Santalum* spp.) and firewood. By the middle of the 19th century, land clearing for large-scale commercial agriculture and *koa* (*Acacia koa*) logging had begun to transform much of the remaining natural vegetation cover at low and middle elevations into one-crop plantations, primarily sugar cane (*Saccharum officinarum*). Unlike the shifting cultivation that characterized the Hawaiians' agricultural use of the upland slopes, modern agriculture completely and permanently removed all native plant cover. The last few decades of the 19th century and the early 20th century saw the conversion of vast tracts of upland forests into cattle ranches and the spread of alien grasses, often leaving very little or no unmodified forest sandwiched between plantations and ranchlands. Land clearing and conversion of natural habitats to agricultural and urban uses have been identified as major causes of extinction of endemic Hawaiian plants and animals (Simon 1987); perhaps 10% of the Hawaiian flora has become extinct and another 40-50% is threatened with extinction (Wagner et al. 1985; Vitousek et al. 1987). Extinction among Hawaiian animals has been even more dramatic; two-thirds of the native bird species, more than 50% of the endemic land snails and a large but unknown percentage of Hawaiian insect species are already extinct (Howarth et al. 1988).

The intentional and accidental introduction of alien plants accelerated at the beginning of the 20th century and resulted in the loss or degradation of most of the remaining natural vegetation of the lowlands and a significant proportion of upland habitats, particularly since the mid-20th century emphasis on multiple use of public forests for feral game animals. By 1980, an estimated 60,730 ha (150,000 a) of Conservation-zoned forest land had been invaded by seriously disruptive alien plants (Hawaii State Department of Land and Natural Resources 1980). Few natural systems are totally immune to the invasion of alien plants, but upper-elevation forests with difficult terrain and no or low populations of feral animals have fared the best. In their analysis of the distribution of six important alien plants in the uplands of Hawai'i Island, Jacobi and Warshauer (in press) found that those ecosystems with the least disturbance from animals or humans were "the most resistant to colonization by noxious introduced plants."

Particularly in dry and mesic vegetation, the widespread invasion and establishment of alien grasses have resulted in a great increase in the size and frequency of fires, which are typically deleterious to native plants. It is likely that some of Hawai'i's alien plant problems are intractable, and many areas currently supporting diverse native vegetation will be overwhelmed as invaders intensify and expand their ranges. This scenario is made even more likely if steps are not taken to control feral animals and prevent fires in important natural areas: those that are nearly intact, have high biological diversity, and contain many rare species.

Mid-20th century attempts to establish a commercial forestry industry in Hawai'i resulted in the loss of native forests and other plant and animal communities on public lands and, perhaps of more far-reaching impact, the introduction of alien plants. In recent years new industrial uses for remote, little-disturbed sites have come to threaten areas previously protected from development by their unsuitability for agriculture or distance from population centers. The trend for expansion of commercial activities and urbanization into remaining native ecosystems in the uplands and undeveloped lowlands will doubtless continue as Hawai'i's population grows and the visitor industry continues to flourish.

After more than a millenium of human occupation, few remnants of natural vegetation may still be found in the coastal and lowland zones of Hawai'i, still the sites of most human occupation and activity. What does remain is often degraded and simplified by alien organisms and continuing destructive land practices. Thus, the few lowland sites still containing a diversity of native organisms are especially precious to those who would preserve examples of the original Hawai'i. More than 75% of the recognized types of plant communities remaining in the coastal and lowland zones are considered to be rare. Although examples of most of these do occur on either State or Federal land, more than 30 of 95 coastal and lowland plant communities are not represented in areas designated for protection (The Nature Conservancy of Hawaii 1987).

The montane and subalpine zones have fared somewhat better than the lowlands, in part because of factors of time and distance. Today, however, with modern developments in technology and access, and continuing problems with the spread and intensification of alien organisms, almost no area is completely free of the possibility of human manipulation. Despite the large areas in these zones still covered by native vegetation, more than 80% of montane communities and all of those in the subalpine zone are considered rare, because of past disturbance and present vulnerability (The Nature Conservancy of Hawaii 1987).

Gagné (1988) estimated that less than 10% of the land on Kaua'i, O'ahu, Maui, and Hawai'i continued to support undisturbed native forest. Of the four Islands he reviewed, heavily urbanized O'ahu showed the lowest percentage of forest cover even when forests disturbed or replaced by nonnative plants were considered. Others have noted the paucity of native vegetation on O'ahu. Mueller-Dombois (1973) observed that the native lowland forests of O'ahu had been replaced by alien species except for one area of dry and mesic forest in the northern Wai'anae Range. Even the wet 'ōhi'a (Metrosideros spp.) forests of the Ko'olau Summit area are declining in cover, resulting in drier conditions and depletion of understory species (Obata 1985b). A recent map shows *'ōhi'a/hāpu'u (Metrosideros polymorpha/Cibotium* spp.*)* forest covering the upper reaches of the Ko'olau Mountains on O'ahu, (Little and Skolmen 1989), but little undisturbed vegetation actually remains there.

Other main Hawaiian Islands (excluding Ni'ihau, Kaho'olawe, and Lāna'i) have fared somewhat better than O'ahu and still have significant amounts of native vegetation. The central portion of Kaua'i, amounting to perhaps one-third of the island, still supports native wet forests (Little and Skolmen 1989); the combined categories of relatively undisturbed forests and those disturbed to some degree were estimated by Gagné (1988) to cover 40% of Kaua'i. The Nature Conservancy of Hawaii (1987) identified and listed a number of intact or nearly intact native communities on the island of Kaua'i, inlcuding wet forests, bogs, and rare types of dry and mesic forests.

Moloka'i, despite its relatively small size and the loss of most of its dry lowland vegetation, still contains a significant amount of nearly "pristine" wet 'ōhi'a forest at upper elevations (Little and Skolmen 1989), as well as important bog communities. The Nature Conservancy Preserves and State Natural Area Reserves contain examples of many of the remaining natural communities on Moloka'i (The Nature Conservancy of Hawaii 1987); these areas have legal protection and are (or will soon be) managed for their natural values.

Maui, the second-largest of the Hawaiian Islands, currently supports large stands of native wet forest (mostly 'ōhi'a) at upper elevations on both its eastern and western sections (Little and Skolmen 1989). East Maui also contains important subalpine and alpine communities (The Nature Conservancy of Hawaii 1987), but large areas of former dry and mesic forests of leeward Haleakalā have been severely degraded (Medeiros et al. 1986; Little and Skolmen 1989). Maui vegetation benefits greatly from the protection afforded to National Parks and State Natural Area Reserves, which contain many of the known natural plant communities of the Island (The Nature Conservancy of Hawaii 1987).

The question of how much relatively undisturbed natural vegetation remains is best understood for the island of Hawai'i, where vegetation data from the U.S. Fish and Wildlife Service Forest Bird Survey have been analyzed and vegetation types mapped by Jacobi (in press b). Jacobi and Scott (1985) reported large expanses (more than 100,000 ha or 247,000 a) of wet 'ōhi'a forest on Hawai'i Island (above 500 m or 1,640 ft elevation) that were dominated by native plants and not severely invaded by alien plants. In contrast, they found that most native dry and mesic vegetation types of the montane and subalpine zones contained significant cover of alien plant species, and in some cases were completely dominated by nonnative plants. The best-preserved dry and mesic communities were those dominated by 'ōhi'a with other native trees and shrubs; more than 50,000 ha (123,500 a) of these vegetation types contained few alien plants. A large area of dry and mesic 'ōhi'a shrublands, including pioneer communities on lava flows, also remained intact; this amounted to more than 35,000 ha (86,450 a) (Jacobi and Scott 1985).

Although much of the natural vegetation and associated animal species of the Hawaiian Islands has been lost, the surviving biota display great resilience, and what remains is of great biological significance (Howarth et al. 1988). Management of remaining natural aras in Hawai'i is possible, albeit expensive, and an added benefit is that knowledge gained in managing highly invaded island ecosystems will ultimately prove useful for conservation of continental systems (Loope et al. 1988). Even those ecosystems that have been damaged or degraded by alien organisms and past land use may have the "capacity for significant recovery" if the agents of disturbance are controlled (Jacobi and Scott 1985).

Considerable area still covered by native vegetation, particularly in high-rainfall regions, owes its continued existence to the early 20th century establishment of forest reserves and watersheds. Many important natural communities are still found in forest reserves, which currently receive little protective management. However, former parts of many forest reserves are now included in the State Natural Area Reserves System, which was established in 1970 to preserve native plants and animals and important geological features. Currently the System includes 18 reserves on five Islands, encompassing more than 43,840 ha (108,288 a). Management plans for these areas have been developed and include fencing and ungulate and alien plant control. One State Wilderness Preserve (Alaka'i) and several sanctuaries also protect important plants, animals, or communities.

Other agencies and organizations that protect natural areas in Hawai'i include the National Park Service, U.S. Fish and Wildlife Service, and The Nature Conservancy of Hawaii. Altogether, these areas designated for protection total more than 197,000 ha (486,620 a) (Holt 1989). Although not all known natural communities and assemblages of species are contained within these legally protected areas, The Nature Conservancy has identified lands that support the unprotected types, and there is hope that these lands will also receive protection in the near future (The Nature Conservancy of Hawaii 1987). The recent development of protected areas managed for their natural values is a positive and extremely important step toward saving the best of what remains of Hawai'i's natural legacy.

Literature Cited

Abbott, I.A. 1977. The influence of the major food crops on the social system of old Hawai'i. *Newsletter, Hawaiian Bot. Soc.* 16(5):78-79.

Allen, H.H. 1936. Indigene versus alien in the New Zealand plant world. *Ecology* 17(2):187-193.

Allen, M.S. 1983. Analysis of archaeobotanical materials. Rept. 14, pp. 384-400 In J.T. Clark and P.V. Kirch (eds.), *Archaeological investigations of the Mudlane-Waimea-Kawaihae Road corridor, island of Hawai'i. An interdisciplinary study of an environmental transect.* Rept. 83-1. Anthropology Dept., B.P. Bishop Museum, Honolulu.

Anderson, S.J., C.P. Stone, and P.K. Higashino. In press. Distribution and spread of alien plants in Kipahulu Valley, Haleakala National Park, above 700 meters. In C.P. Stone, C.W. Smith, and J.T. Tunison (eds.), *Alien plant invasions in native ecosystems of Hawai'i: management and research.* Univ. Hawaii Coop. Natl. Park Resour. Stud. Unit. Univ. Hawaii Press, Honolulu.

Animal Species Advisory Commission. 1974. *Reviews of the five-year forest planting plan for the State of Hawaii, fiscal years 1972-1976 and the attendant Environmental Impact Statement.* Hawaii State Dept. Land Nat. Resour., Honolulu. 69 pp. + attach.

Anonymous. 1856. The influence of the cattle on the climate of Waimea and Kawaihae, Hawaii. *Sandwich Islands' Monthly Magazine*, Feb.

Anonymous. 1970. Extinction threatens 2,000 isle species. *Honolulu Star-Bulletin*, Thurs., May 14, p. G-14.

Anonymous. 1973. Timber problem: investment big, profit doubtful. *West Hawaii Today*, Thurs., Dec. 20, pp. 34-35.

Anonymous. 1976. *Natural hazards on the island of Hawaii.* U.S. Govt. Printing Office, Washington, D.C.

Anonymous. 1986. Species-specific fungus approved for use against Coster's curse. *'Elepaio* 46(18):194-195.

Anonymous. 1987. Marijuana growing on conservation district lands. *'Elepaio* 47(6):69-70.

Anonymous. 1988. Forest and watershed - Interim Workgroup report to Speaker Daniel J. Kihano: November 25, 1988. Unpubl. rept. Research Div., Hawaii Volcanoes Natl. Park. 16 pp.

Anonymous. 1989a. Exotic "Brentwood" tree available at plant sales. *Hawaii Tribune-Herald*, Thurs., Feb. 9, p. 13.

Anonymous. 1989b. 'Alala (Hawaiian crow) recovery; Audubon Hawai'i calls for immediate government action. *Greenprint (Audubon Conservation News)* 1(2):1-2.

Anonymous. 1989c. Mullein discovered in Haleakala National Park. *Newsletter, Hawaiian Bot. Soc.* 25(3):89.

Apple, R.A. 1954. A history of land acquisition for Hawaii National Park to December 31, 1950. M.A. Thesis, Univ. Hawaii, Honolulu. 158 pp.

Apple, R.A. 1971. Koa and lehua timber harvesting and product utilization; religio-ecological relationships in Hawaii, A.D. 1778. National Park Service and B.P. Bishop Museum. Unpubl. ms. Library, Hawaii Volcanoes Natl. Park. 43 pp.

Armstrong, F.E. 1937. *A survey of small farming in Hawaii*. Univ. Hawaii Research Publ. No. 14, Honolulu. 90 pp.

Armstrong, R.W. (ed.). 1983. *Atlas of Hawaii*, 2nd ed. Geography Dept., Univ. Hawaii. Univ. Press Hawaii, Honolulu. 238 pp.

Atkinson, I.A.E. 1970. Successional trends in the coastal and lowland forest of Mauna Loa and Kilauea Volcanoes, Hawaii. *Pac. Sci.* 24:387-400.

Atkinson, I.A.E. 1977. A reassessment of factors, particularly *Rattus rattus* L., that influenced the decline of endemic forest birds in the Hawaiian Islands. *Pac. Sci.* 31(2):109-133.

Atkinson, I.A.E. 1985. The spread of commensal species of *Rattus* to oceanic islands and their effects on island avifaunas. pp. 35-81 In P.J. Moors (ed.), *Conservation of island birds*. Internatl. Council Bird Preserv. Tech. Publ. 3. Cambridge, England.

Baker, H.L. 1961. The land situation in the state of Hawaii. Land Study Bureau Circ. No. 13, Univ. Hawaii, Honolulu. 28 pp.

Baker, J.K. 1979. Studies on the ecological impact of introduced roof rats upon native flora in Hawaii Volcanoes National Park. pp. 461-466 In C.W. Smith (ed.), *Proc. 2nd Conf. Nat. Sci., Hawaii Volcanoes Natl. Park*. Univ. Hawaii Coop. Natl. Park Resour. Stud. Unit. Botany Dept., Univ. Hawaii, Honolulu.

Baker, J.K., and S. Allen. 1976. Studies on the endemic Hawaiian genus *Hibiscadelphus (hau-kuahiwi)*. pp. 19-22 In C.W. Smith (ed.), *Proc. 1st Conf. Nat. Sci., Hawaii Volcanoes Natl. Park*. Univ. Hawaii Coop. Natl. Park Resour. Stud. Unit. Botany Dept., Univ. Hawaii, Honolulu.

Baldwin, P.H., and G.O. Fagerlund. 1943. The effect of cattle grazing on koa reproduction in Hawaii National Park. *Ecology* 24:118-122.

Barrera, W., Jr., and M. Kelly. 1974. *Archaeological and historical surveys of the Waimea to Kawaihae Road corridor, island of Hawai'i*. Hawaii Historic Preservation Rept. 74-1. Anthropology Dept., B.P. Bishop Museum, Honolulu. 84 pp.

Bartlett, H.H. 1962. Fire, primitive agriculture, and grazing in the tropics. pp. 692-715 In W.L. Thomas, Jr. (ed.), *Man's role in changing the face of the earth*. Univ. Chicago Press, Chicago, Illinois.

Beaglehole, J. (ed.). 1967. *The journals of Captain James Cook on his voyages of discovery. The voyage of the Resolution and Discovery 1776-1780*. Part One. J.C. Beaglehole (ed.) Univ. Press, Cambridge, England. 718 pp.

Becking, J.H. 1979. Nitrogen fixation by *Rubus ellipticus* J.E. Smith. *Plant and Soil* 53:541-545.

Benson, B. 1970. Scientists differ on isle forest policy. *The Honolulu Advertiser*, Aug. 21.

Bird, I.L. 1966. *Six months in the Sandwich Islands*. Univ. Hawaii Press, Honolulu. 278 pp.

Bishop, D.H. 1989. Wailuku project gets board's OK. *Hawaii Tribune-Herald*, Sun., Sept. 24, pp. 1, 8.

Bloxam, A. 1925. *Diary of Andrew Bloxam, naturalist of the "Blonde" on her trip from England to the Hawaiian Islands 1824-25*. B.P. Bishop Museum Spec. Publ. 10. Bishop Mus. Press, Honolulu.

Blumenstock, D.I., and S. Price. 1967. *Climates of the states: Hawaii*. Environ. Sci. Data Service, Climatology of the United States, No. 60-51. U.S. Govt. Printing Office, Washington, D.C. 27 pp.

Brewbaker, J.L., and B.K. Styles (eds.). 1984. Economically important nitrogen fixing tree species planted in Hawaii. *Newsletter, Hawaiian Bot. Soc.* 23:30-35.

Bryan, L.W. 1926a. Letter to Albert Duvel, Dec. 8. L.W. Bryan Forestry Collection, B.P. Bishop Museum, Honolulu. 1 p.

Bryan, L.W. 1926b. Letter to the *Hilo Tribune Herald*, May 13. (Publ. May 14). L.W. Bryan Forestry Collection, B.P. Bishop Museum, Honolulu. 2 pp.

Bryan, L.W. 1926-27. Hawaii Progress Reports. Unpubl. ms. L.W. Bryan Forestry Collection, B.P. Bishop Museum, Honolulu.

Bryan, L.W. 1931. Hawaii's wild goat problem. *Paradise Pac.* 43(12):77-78.

Bryan, L.W. 1937. Wild sheep in Hawaii. *Pac. Discovery* 49:19, 31.

Bryan, L.W. 1944a. Letter to H.L. Lyon, Aug. 3. L.W. Bryan Forestry Collection, B.P. Bishop Museum, Honolulu. 1 p.

Bryan, L.W. 1944b. Letter to H.L. Lyon, Sept. 12. L.W. Bryan Forestry Collection, B.P. Bishop Museum, Honolulu. 1 p.

Bryan, L.W. 1947. Twenty-five years of forestry work on the island of Hawaii. *The Hawaiian Planters' Record* 51(1):1-80.

Bryan, L.W. 1954. Tropical black raspberry *(Rubus albescens)* "a new plant immigrant." Letter to "Hawaii Farmer," Honolulu, Nov. 29, 1954. L.W. Bryan Forestry Collection, B.P. Bishop Museum, Honolulu. 1 pp.

Bryan, L.W. 1957. Final report, lands of Kahaualea, Puna, Hawaii. Prepared for Estate of James Campbell. L.W. Bryan Forestry Collection, B.P. Bishop Museum, Honolulu. 40 pp.

Bryan, L.W. 1961a. History of Hawaiian forestry up to and including 1920. Unpubl. ms. L.W. Bryan Forestry Collection, B.P. Bishop Museum, Honolulu. 17 pp.

Bryan, L.W. 1961b. Report to C. Brewer. History of Hilo Forest Reserve. Unpubl. ms. L.W. Bryan Forestry Collection, B.P. Bishop Museum, Honolulu. 3 pp.

Bryan, L.W. 1966. Afforestation in the Hawaiian Islands, State of Hawaii, U.S.A. Sexto Congreso Forestal Mundial, Madrid, Junio 1966. Unpubl. ms. L.W. Bryan Forestry Collection, B.P. Bishop Museum, Honolulu. 4 pp.

Bryan, L.W. 1977a. Bill Bryan views 50 years of forestry. *Hawaii Tribune-Herald*, Feb. 27. L.W. Bryan Forestry Collection, B.P. Bishop Museum, Honolulu. pp. C 14-16.

Bryan, L.W. 1977b. Letter to the Editor. *Newsletter, Hawaiian Bot. Soc.* 16:63.

Bryan, L.W. 1980. Reminiscences of Lester W. Bryan. pp. 19-27 In C.E. Hartt (ed.), *Harold Lloyd Lyon, Hawaiian sugar botanist*. Harold L. Lyon Arboretum, Univ. Hawaii, Honolulu.

Bryan, L.W. n.d. Forestry on the plantation. Unpubl. ms. L.W. Bryan Forestry Collection, B.P. Bishop Museum, Honolulu. 6 pp.

Bryan, W.A. 1915. *Natural history of Hawaii*. Hawaiian Gazette, Honolulu. 596 pp.

Buck, M.G. 1982. *Hawaiian treefern harvesting affects forest regeneration and plant succession*. Research Note PSW-355. Pac. SW Forest and Range Expt. Sta. U.S. Forest Service, Berkeley, Calif. 8 pp.

Buck, M.G., and R.H. Imoto. 1982. *Growth of 11 introduced species on selected forest sites in Hawaii*. Research Paper PSW-169. Pac. SW Forest and Range Expt. Sta. U.S. Forest Service, Berkeley, Calif. 11 pp.

Buck, M.G., P.G. Costales, and E.Q.P. Petteys. 1979. *Forest plantation survey, Biomass Energy Tree Farm Program*. Hawaii State Dept. Land Nat. Resour., Div. Forestry, Honolulu. 22 pp.

Burr, T.A. 1984. Introduced birds in Hawaii and some associated state management problems. pp. 92-97. *Proc. 24th Annual Forestry and Wildlife Conf.*, Honolulu. 164 pp.

Burton, P.J. 1980. Plant invasion into an 'ohi'a-treefern rain forest following experimental canopy opening. pp. 21-39 In C.W. Smith (ed.), *Proc. 3rd. Conf. Nat. Sci., Hawaii Volcanoes Natl. Park*. Univ. Hawaii Coop. Natl. Park Resour. Stud. Unit. Botany Dept., Univ. Hawaii, Honolulu.

Campbell, D.H. 1920. Some botanical and environmental aspects of Hawaii. *Ecology* 1:257-269.

Campbell, D.J. 1978. The effects of rats on vegetation. pp. 99-100 In P.R. Dingwall, I.A.E. Atkinson, and C. Hay (eds.), *The ecology and control of rodents in New Zealand native reserves*. New Zealand Dept. Lands Surv. Info. Ser. 4.

Canfield, J.E. 1986. The role of edaphic factors and plant water relations in palnt distribution in the bog/wet forest complex of Alaka'i Swamp, Kaua'i, Hawai'i. PhD Diss., Botany Dept., Univ. Hawaii, Honolulu. 280 pp.

Carlquist, S. 1980. *Hawaii: a natural history. Geology, climate, native flora and fauna above the shoreline*. Pac. Trop. Bot. Gdn., Lawai, Kauai, Hawaii. 468 pp.

Carlson, N.K., and L.W. Bryan. 1963. The Honaunau Forest: an appraisal after seven years of planting. *J. Forestry* 61(9):643-647.

Carlson, N.K., and L.W. Bryan. n.d. A report on the Honaunau Forest Reserve, S. Kona, Hawaii, its present condition and its potential. Unpubl. rept. L.W. Bryan Forestry Collection, B.P. Bishop Museum, Honolulu. 103 pp.

Carr, G.D., E.A. Powell, and D.W. Kyhos. 1986. Self-incompatability in the Hawaiian Madiinae (Compositae): an exception to Baker's rule. *Evolution* 40(2):430-434.

Carson, H.L. Fluctuations in size of certain *Drosophila* populations in the ʻOlaʻa tract, Hawaii Volcanoes National Park. p. 40 In C.W. Smith (ed.), *Proc. 4th Conf. Nat. Sci., Hawaii Volcanoes Natl. Park.* Univ. Hawaii Coop. Natl. Park Resour. Stud. Unit. Botany Dept., Univ. Hawaii, Honolulu.

Chai, D.K., L.W. Cuddihy, and C.P. Stone. In press. *An inventory and assessment of anchialine pools in Hawaii Volcanoes National Park from Wahaʻula to Kaʻaha, Puna and Kaʻu, Hawaiʻi.* Univ. Hawaii Coop. Natl. Park Resour. Stud. Unit Tech. Rept. 69. Botany Dept., Univ. Hawaii, Honolulu. 37 pp.

Chamber of Commerce of Hilo. 1926. Resolution concerning destruction by fire of the Panaewa Forest Reserve. May 11, 1926. L.W. Bryan Forestry Collection, B.P. Bishop Museum, Honolulu. 3 pp.

Chapman, P.S., and P.V. Kirch. 1979. *Archeological excavations at seven sites, southeast Maui, Hawaiian Islands.* Rept. 79-1. Anthropology Dept., B.P. Bishop Museum, Honolulu.

Char, W.P., and N. Balakrishnan. 1979. Ewa Plains botanical survey. U.S. Fish and Wildl. Service, Honolulu. Unpubl. rept. Research Div., Hawaii Volcanoes Natl. Park.

Char, W.P., and C.H. Lamoureux. 1985. *Puna geothermal area biotic assessment, Puna District, County of Hawaii.* Prepared for Hawaii State Dept. Planning and Econ. Dev. Botany Dept., Univ. Hawaii, Honolulu. 127 pp. + appendices.

Chow, W.T. 1983. Urbanization: six propositions. pp. 167-185 In J.R. Morgan (ed.), *Hawaii, a geography.* Westview Press, Boulder, Colorado.

Christensen, C.C. 1983. Analysis of land snails. Rept. 17, pp. 449-471 In J.T. Clark and P.V. Kirch (eds.), *Archaeological investigations of the Mudlane-Waimea-Kawaihae Road corridor, island of Hawaiʻi. An interdisciplinary study of an environmental transect.* Rept. 83-1. Anthropology Dept., B.P. Bishop Museum, Honolulu.

Christensen, C.C., and P.V. Kirch. 1981. Land snails and environmental change at Barbers Point, Oahu, Hawaii. Abstract. *Bull. Amer. Malacol. Union* 1981:31.

Christophersen, E., and E.L. Caum. 1931. *Vascular plants of the Leeward Islands, Hawaii.* B.P. Bishop Museum Bull. 81. Honolulu. 41 pp.

Clark, H. 1982. Botanist says rare fern can survive geothermal project. *The Honolulu Advertiser*, Tues., Nov. 16, p. A-7.

Clark, J.T. 1983a. Archaeological investigations of agricultural sites in the Waimea area. Rept. 8, pp. 293-314 In J.T. Clark and P.V. Kirch (eds.), *Archaeological investigations of the Mudlane-Waimea-Kawaihae Road corridor, island of Hawaiʻi. An interdisciplinary study of an environmental transect.* Rept. 83-1. Anthropology Dept., B.P. Bishop Museum, Honolulu.

Clark, J.T. 1983b. Archaeological investigations in section 4. Rept. 7, pp. 240-292 In J.T. Clark and P.V. Kirch (eds.), *Archaeological investigations of the Mudlane-Waimea-Kawaihae Road corridor, island of Hawaiʻi. An interdisciplinary study of an environmental transect.* Rept. 83-1. Anthropology Dept., B.P. Bishop Museum, Honolulu.

Clark, J.T., and P.V. Kirch (eds.). 1983. *Archaeological investigations of the Mudlane-Waimea-Kawaihae Road corridor, island of Hawaiʻi. An interdisciplinary study of an environmental transect.* Rept. 83-1. Anthropology Dept., B.P. Bishop Museum, Honolulu. 532 pp.

Clarke, G., L. Cuddihy, J. Davis, and S. Anderson. 1981. A botanical reconnaissance of Malama-Ki Forest Reserve, Hawaii. Unpubl. rept. Hawaii State Dept. Land Nat. Resour., Div. Forestry, Hilo, Hawaii. 19 pp.

Clarke, G.G., L.W. Cuddihy, J.A. Davis, and S.J. Anderson. 1983. A botanical survey of Keauhou Ranch and Kilauea Forest, Hawaii, with emphasis on the endangered plant species *Vicia menziesii* Spreng. Unpubl. rept. Hawaii State Dept. Land Nat. Resour., Div. Forestry and Wildlife, Hilo, Hawaii. 219 pp.

Clarke, G., L. Cuddihy, J. Davis, and R. Kubo. 1980. Honaunau koa reforestation site vegetation survey (62 acres). Unpubl. rept. Hawaii State Dept. Land Nat. Resour., Div. Forestry, Hilo, Hawaii. 13 pp. + appendices.

Clausen, C.P. (ed.). 1978. *Introduced parasites and predators of arthropod pests and weeds: a world view.* Agric. Handbk. 480. Agric. Research Service, U.S. Dept. Agric., Washington, D.C. 545 pp.

Cole, F.R., L.L. Loope, and A.C. Medeiros. 1986a. Population biology and food habits of introduced rodents in high-elevation shrubland of Haleakala Park, Maui, Hawaii. (Abstract.) p. 220, Program, 4th Internatl. Congress of Ecology, Syracuse, N.Y., August 10-16, 1986.

Cole, F.R., L.L. Loope, and A.C. Medeiros. 1986b. The effects of introduced gamebirds on the biota of the high elevation shrubland of Haleakala National Park, Maui, Hawaii. (Abstract.) p. 102, Program, 4th Internatl. Congress of Ecology, Syracuse, N.Y., August 10-16, 1986.

Cole, F.R., A.C. Medeiros, and L.L. Loope. 1986c. The impact of Argentine ants *(Iridomyrmex humilis)* on the arthropod fauna of Haleakala National Park, Maui, Hawaii. (Abstract.) p. 282, Program, 4th Internatl. Congress of Ecology, Syracuse, N.Y., August 10-16, 1986.

Conant, S. 1980. *Birds of the Kalapana Extension.* Univ. Hawaii Coop. Nat. Park Resour. Stud. Unit Tech. Rept. 36. Botany Dept., Univ. Hawaii, Honolulu. 43 pp.

Conant, S. 1985. Recent observations on the plants of Nihoa Island, Northwestern Hawaiian Islands. *Pac. Sci.* 39(2):135-149.

Conant, S. 1986. No Na Leo 'Ole. *'Elepaio* 46(8):86.

Conant, S., C.C. Christensen, P. Conant, W.C. Gagné, and M.L. Goff. 1984. The unique terrestrial biota of the Northwestern Hawaiian Islands. *Proc. 2nd Symp. Resour. Invert. Northwest. Hawaiian Islands, May 25-27, 1983.* Vol. 1:77-94.

Cooper, G., and G. Daws. 1985. *Land and power in Hawaii, the Democratic years.* Benchmark Books, Honolulu. 518 pp.

Cooray, R.G. 1974. Stand structure of a montane rain forest on Mauna Loa, Hawaii. Master's Thesis, Botany Dept., Univ. Hawaii, Honolulu. 165 pp.

Cooray, R.G., and D. Mueller-Dombois. 1981. Feral pig activity. pp. 309-317 In D. Mueller-Dombois, K.W. Bridges, and H.L. Carson (eds.), *Island ecosystems: biological organization in selected Hawaiian communities.* Hutchinson Ross Publ. Co., Stroudsburg, Pennsylvania.

Coulter, J.W. 1931. *Population and utilization of land and sea in Hawaii 1853.* B.P. Bishop Museum Bull. 88. Honolulu. 33 pp.

Corn, C., D. Herbst, and P. Higashino. 1978. Report of the botanical survey on the 200-acre reforestation site at Keauhou-Bishop Estate tract for *Vicia menziesii*. Unpubl. rept. U.S. Fish and Wildlife Service, Honolulu. 4 pp.

Corn, C., W. Char, G. Clarke, and L. Cuddihy. 1980. Kahoʻolawe botanical survey. Unpubl. rept. Hawaii State Dept. Land Nat. Resour., Div. Forestry and Wildlife, Honolulu. 32 pp.

Corn, C., G. Clarke, L. Cuddihy, and L. Yoshida. 1979. A botanical reconnaissance of Kalalau, Honopu, ʻAwaʻawapuhi, Nualolo, and Miloliʻi Valleys and shorelines -- Na Pali, Kauaʻi. Unpubl. rept. Hawaii State Dept. Land Nat. Resour., Div. Forestry, Hilo, Hawaii. 9 pp. + appendices.

Creighton, T.H. 1978. *The lands of Hawaii: their use and misuse*. Univ. Press Hawaii, Honolulu. 417 pp.

Critchlow, L. 1980. Trees: forum focuses on future of forests. Kona Coast spotlight. *Hawaii Tribune-Herald*, Mar. 28, pp. 9-10.

Critchlow, L. 1988. Moths a new weapon in fight against spread of nasty weed. *Hawaii Tribune-Herald*, Thurs., Dec. 1. pp. 1, 14, 16.

Cuddihy, L.W. 1984. *Effects of cattle grazing on the mountain parkland ecosystem, Mauna Loa, Hawaii*. Univ. Hawaii Coop. Nat. Park Resour. Stud. Unit Tech. Rept. 51. Botany Dept., Univ. Hawaii, Honolulu. 135 pp.

Cuddihy, L.W. 1989. Vegetation zones of the Hawaiian Islands. pp. 27-37 In C.P. Stone and D.B. Stone (eds.), *Conservation biology in Hawaiʻi*. Univ. Hawaii Coop. Natl. Park Resour. Stud. Unit. Univ. Hawaii Press, Honolulu.

Cuddihy, L.W., J.A. Davis, and S.J. Anderson. 1982a. A botanical survey of twelve cinder cones in South Kohala, island of Hawaiʻi. Unpubl. rept. Hawaii State Dept. Land Nat. Resour., Div. Forestry and Wildlife, Endangered Plant Species Program, Hilo, Hawaii. 61 pp.

Cuddihy, L.W., J.A. Davis, and S.J. Anderson. 1982b. A botanical reconnaissance of the proposed Kamakou Preserve, Molokaʻi (2774 acres). Unpubl. rept. Hawaii State Dept. Land Nat. Resour., Div. Forestry and Wildlife, Endangered Plant Species Program, Hilo, Hawaii. 95 pp.

Cuddihy, L.W., C.P. Stone, and J.T. Tunison. 1988. Alien plants and their management in Hawaii Volcanoes National Park. *1988 Trans. Western Sect. Wildlife Soc.* 24:42-46.

Culliney. J.L. 1988. *Islands in a far sea: nature and man in Hawaiʻi*. Sierra Club Books, San Francisco, Calif. 410 pp.

Davidson, J.M. 1979. New Zealand. pp. 222-248 In J.D. Jennings (ed.), *The prehistory of Polynesia*. Harvard Univ. Press, Cambridge, Mass. and London, England.

Davis, B.D. 1982. Horticultural adaptation and ecological change in southwestern Oʻahu: preliminary evidence from Barbers Point. pp. 51-59 In C.W. Smith (ed.), *Proc. 4th Conf. Nat. Sci., Hawaii Volcanoes Natl. Park*. Univ. Hawaii Coop. Natl. Park Resour. Stud. Unit. Botany Dept., Univ. Hawaii, Honolulu.

Davis, J. 1987. *Miconia calvescens* on Hawaii. *Newsletter, Hawaiian Bot. Soc.* 26(2):32.

Degener, O. 1934, 1936, 1938. *Flora Hawaiiensis. The new illustrated flora of the Hawaiian Islands*. Privately published, Honolulu.

Degener, O., and I. Degener. 1974. Appraisal of Hawaiian taxonomy. *Phytologia* 29(3):240-246.

Diong, C.H. 1982. Population biology and management of the feral pig (*Sus scrofa* L.) in Kipahulu Valley, Maui. PhD Thesis, Univ. Hawaii, Honolulu. 408 pp.

Doerr, J.E., Jr. 1931. Blackberry. *Nature notes, Hawaii National Park* 1(1):8. Library, Hawaii Volcanoes Natl. Park.

Doerr, J.E., Jr. 1932. Pulu. *Nature notes, Hawaii National Park* 2(2):9-16. Library, Hawaii Volcanoes Natl. Park.

Dorsett, E.L. 1954. Hawaiian whaling days. *The Amer. Neptune* 14(1):42-46.

Douglas, D. 1914. *Journal kept by David Douglas during his travels in North America 1823-1827.* William Wesley & Son, London. 364 pp.

Dye, T. In press. Tales of two cultures: traditional historical and archaeological interpretations of Hawaiian prehistory. *Occas. Papers Bernice P. Bishop Museum.*

Eckholm, E. 1979. Forest renewal in India. *Nat. Hist.* 88(6):12-27. Cited by Kelly 1983.

Egler, F.E. 1939. Vegetation zones of Oahu, Hawaii. *Empire Forest J.* 18:44-57.

Egler, F.E. 1942. Indigene versus alien in the development of arid Hawaiian vegetation. *Ecology* 23(1):14-23.

Ellis, W. 1969. *Polynesian researches: Hawaii. Journal of William Ellis.* Charles E. Tuttle Co., Rutland Vermont and Tokyo, Japan. 471 pp.

Ellis, W. 1827 [Reprinted 1963]. *Journal of William Ellis. Narrative of a tour of Hawai'i or Owhyhee; with remarks on the history, traditions, manners, customs, and language of the inhabitants of the Sandwich Islands.* Advertiser Publ. Co., Ltd., Honolulu. 342 pp.

Emory, K.P., W.J. Bonk, and Y.H. Sinoto. 1969. *Waiahukini shelter, Site H8, Ka'u, Hawaii.* Pac. Anthropological Records 7. Anthropology Dept., B.P. Bishop Museum, Honolulu. 12 pp.

Ewel, J.J., D.S. Ojima, D.A. Karl, and W.F. DeBusk. 1982. *Schinus in successional ecosystems of Everglades National Park.* Rept. T-676, U.S. Natl. Park Service, S. Fla. Res. Center, Everglades Natl. Park, Homestead, Florida. 141 pp.

Fagerlund, G.O. 1943. Herbarium label for *Hedychium gardnerianum.* Hawaii National Park Quarters No. 1. August 11, 1943. Herbarium, Research Div., Hawaii Volcanoes Natl. Park.

Fagerlund, G.O. 1947. The exotic plants of Hawaii National Park. *Natural History Bull.* No. 10. U.S. Dept. Interior, National Park Service, Hawaii National Park. 62 pp.

Fernwood Industries of Hawaii. 1971. Hapu'u Harvesting Plan, land of Kahaualea -- island of Hawaii. L.W. Bryan Forestry Collection, B.P. Bishop Museum, Honolulu. 13 pp.

Fisher, H.I. 1948. The question of avian introductions in Hawaii. *Pac. Sci.* 2:59-64.

Fisher, T. 1973. *Hawaii: growing pains in paradise.* Vol. 29, No. 3. Population Ref. Bureau, Inc. 40 pp.

Fosberg, F.R. 1948a. Derivation of the flora of the Hawaiian Islands. pp. 107-119 In E.C. Zimmerman, *Insects of Hawaii*, Vol. 1. Univ. Press Hawaii, Honolulu.

Fosberg, F.R. 1948b. Immigrant plants in the Hawaiian Islands II. *Univ. Hawaii Occas. Papers* 46:3-17.

Fosberg, F.R. 1966. Vascular plants. pp. 153-238 In M.S. Doty and D. Mueller-Dombois, *Atlas for bioecology studies in Hawaii Volcanoes National Park*. Hawaii Bot. Sci. Paper No. 2, Univ. Hawaii, Honolulu.

Fosberg, F.R. 1972. *Guide to Excursion III, Tenth Pacific Science Congress*. Botany Dept., Univ. Hawaii, with assistance from Hawn. Bot. Gdns. Fndtn., Inc., Honolulu. 249 pp.

Frierson, B.B. 1973. A study of land use and vegetation change: Honouliuli, 1790-1925. Unpubl. rept. Hawaiian Collection, Hamilton Library, Univ. Hawaii, Honolulu.

Fujioka, F.M., and D.M. Fujii. 1980. *Physical characteristics of selected fine fuels in Hawaii -- some refinements on surface area-to-volume calculations*. Res. Note PSW-348. Pac. SW Forest and Range Expt. Sta., U.S. Forest Service, Berkeley, Calif. 7 pp.

Funk, E. 1978. Hawaiian fiber plants. *Newsletter, Hawaiian Bot. Soc.* 17:27-35.

Gage, R.P. II. 1988. A note on the Cook voyage collection in Berne Historical Museum. *Archaeology on Kaua'i* (Anthropology Club of Kaua'i) 15(1):1-7.

Gagné, W.C. 1974. Notes on the present status of the native Hawaiian flora. *Newsletter, Hawaiian Bot. Soc.* 13(5):19-20.

Gagné, W.C. 1979. Canopy-associated arthropods in *Acacia koa* and *Metrosideros* tree communities along an altitudinal transect on Hawaii Island. *Pac. Insects* 21(1):56-82.

Gagné, W.C. 1981. Status of Hawaii's endangered species: insects and land snails. *'Elepaio* 42(4):31-36.

Gagné, W.C. 1988. Conservation priorities in Hawaiian natural systems. *Bioscience* 38(4):264-270.

Gagné, W.C., and C.C. Christensen. 1985. Conservation status of native terrestrial invertebrates in Hawai'i. pp. 105-126 In C.P. Stone and J.M. Scott (eds.), *Hawai'i's terrestrial ecosystems: preservation and management*. Univ. Hawaii Coop. Natl. Park Resour. Stud. Unit. Univ. Hawaii Press, Honolulu.

Gagné, W.C., and L.W. Cuddihy. In press. Vegetation. In W.L. Wagner, D.R. Herbst, and S.H. Sohmer, *Manual of the flowering plants of Hawai'i*. Bishop Museum and Univ. Hawaii Presses.

Gagné, W.C., and S.L. Montgomery. 1986. H-3 impacted by wildlife resources. *'Elepaio* 46(4):158-159.

Gambino, P., A.C. Medeiros, and L.L. Loope. 1987. Introduced vespids *Paravespula pensylvanica* prey on Maui's endemic arthropod fauna. *J. Trop. Ecol.* 3:169-170.

Gardner, D.E. 1980. An evaluation of herbicidal methods of strawberry guava control in Kipahulu Valley. pp. 63-69 In C.W. Smith (ed.), *Resources base inventory of Kipahulu Valley below 2000 feet, Maui, Hawaii*. Sponsored by The Nature Conservancy. Honolulu.

Gardner, D.E., and C.J. Davis. 1982. *The prospects for biological control of nonnative plants in Hawaiian national parks*. Univ. Hawaii Coop. Nat. Park Resour. Stud. Unit Tech. Rept. 45. Botany Dept., Univ. Hawaii, Honolulu. 55 pp.

Giffard, W.M. 1918. An appeal for action on forestry work. *The Hawn. Planters' Record* 18(6):539-543.

Giffin, J. 1978. *Ecology of the feral pig on the island of Hawaii*. Final Rept., Pittman-Robertson Project No. W-15-3, Study No. 11, 1968-1972. Hawaii State Dept. Land Nat. Resour., Div. Fish and Game. 122 pp.

Giffin, J.G. 1982. Ecology of the mouflon sheep on Mauna Kea. Final Rept., Hawaii Dept. Land Nat. Resour., Div. Forestry and Wildlife. Proj. W-17-R, Study No. R-3. Honolulu. 65 pp. Mimeo.

Green, R.C. 1971. The chronology and age of sites at South Point, Hawaii. *Archeol. and Physical Anthropol. in Oceania* 6(2):170-176.

Griffin, P.B., T. Riley, P. Rosendahl, and H.D. Tuggle. 1971. Archaeology of Halawa and Lapakahi: windward valley and leeward slope. *New Zealand Archeol. Assn. Newsletter* 14:101-112.

Hadley, T.H. 1966. *Waimea Canyon and Kokee, a nature guide*. Hui o Laka. Kauai Publ. Co., Lihue, Kauai, Hawaii. 71 pp.

Hall, W.L. 1904. *The forests of the Hawaiian Islands*. Bull. 48. U.S. Dept. Agric., Bureau Forestry. U.S. Govt. Printing Office, Washington, D.C. 29 pp.

Hammatt, H.H., M.J. Tomonari-Tuggle, and C.F. Streck. 1978. *Archeological investigations at Ha'ena State Park, Halele'a, Kaua'i Island*. Phase II. Archaeological Research Center, Hawai'i, for Hawaii State Dept. Land Nat. Resour./State Parks. 307 pp.

Hatheway, W.H. 1952. Composition of certain native dry forests: Mokuleia, Oahu, T.H. *Ecol. Monogr.* 22:153-168.

Handy, E.S.C., and E.G. Handy. 1972. *Native planters in old Hawaii: their life, lore, and environment*. B.P. Bishop Museum Bull. 233. Honolulu. 641 pp.

Harada-Stone, D. 1989. Riviera developer: vast project doesn't reflect bad planning. *Hawaii Tribune-Herald*, Thurs., Mar. 23, pp. A-1, A-12.

Hartt, C.E. (ed.). 1980. *Harold Lloyd Lyon, Hawaiian sugar botanist*. Harold L. Lyon Arboretum, Univ. Hawaii, Honolulu. 110 pp.

Haselwood, E.L., and G.G. Motter (eds.). 1976. *Handbook of Hawaiian weeds*. Lyon Arboretum Assn., with permission from Expt. Sta./ Hawn. Sugar Planters' Assn., Honolulu. 479 pp.

Hawaii Agricultural Reporting Service. 1983. *Statistics of Hawaiian agriculture*. Hawaii State Dept. Agric., Marketing and Consumer Services Div., and U.S. Dept. Agric., Statistical Reporting Service, Honolulu. 100 pp.

Hawaii Island Economic Development Board. 1988. [Focus on Agriculture]. Biomass is Hawaii's most versatile renewable energy source. Public Service Rept. *Hawaii Tribune-Herald*, Sun., Nov. 13, p. 17.

Hawaii Island Economic Development Board. 1989. [Focus on Agriculture]. Working to preserve native forests. Public Service Rept, Agric. Committee. *Hawaii Tribune-Herald*, Sun., Mar. 5, p. 31.

Hawaii State Department of Agriculture. 1962. Noxious weeds of Hawaii. Div. Plant Industry, Honolulu. Mimeo. 89 pp.

Hawaii State Department of Agriculture. 1978. List of plant species designated as noxious weeds for eradication and control purposes by the Hawaii Department of Agriculture. Div. Plant Industry, Honolulu. Mimeo. 4 pp.

Hawaii State Department of Business and Economic Development. 1988. *The State of Hawaii Data Book, a statistical abstract -- 1988*. Honolulu. 694 pp.

Hawaii State Department of Land and Natural Resources. 1973. *Environmental impact statement for the proposed reforestation project within portions of the Waiakea, Upper Waiakea, and Olaa Forest Reserves*. Div. Forestry. Honolulu. 11 pp. + figures.

Hawaii State Department of Land and Natural Resources. 1974. Forest planting plan, 1972-1976, Division of Forestry. pp. 41-47 In The Animal Species Advisory Commission 1974, *Reviews of the five year forest planting plan for the State of Hawaii, final years 1972-1976 and the attendant Environmental Impact Statement*.

Hawaii State Department of Land and Natural Resources. 1980. *State conservation lands plan and technical reference document*. Honolulu. 232 pp. + append.

Hawaii State Department of Land and Natural Resources. 1987. *State of Hawaii Department of Land and Natural Resources Report to the Governor 1986-87*. Vol. II. Honolulu. 97 pp.

Hawaii State Department of Land and Natural Resources and Department of Planning and Economic Development. 1976. *Forestry potentials for Hawaii*. U.S. Forest Service Region 5, Honolulu. 68 pp.

Hawaii State Department of Planning and Economic Development. 1982. Geothermal power development in Hawaii. Vol. II. *Infrastructure and community services requirements, island of Hawaii*. Prepared for U.S. Dept. Energy. Honolulu.

Hawaii State Department of Planning and Economic Development. 1986. *Geothermal resource subzone designations in Hawaii*. Honolulu. 158 pp. + appendices.

Hawaii State Department of Planning and Economic Development. 1985. *The state of Hawaii data book -- Nov. 1985*. Honolulu. 662 pp.

Hawaii Sugar Planters' Association. 1988. *Hawaiian sugar manual -- 1988*. Honolulu. 26 pp.

Henke, L.A. 1929. *A survey of livestock in Hawaii*. Univ. Hawaii Research Publ. No. 5. Honolulu. 82 pp.

Herbst, D.R., and W.L. Wagner. In press. Alien plants on the Northwestern Hawaiian Islands. In C.P. Stone, C.W. Smith, and J.T. Tunison (eds.), *Alien plant invasions in native ecosystems of Hawai'i: management and research*. Univ. Hawaii Coop. Natl. Park Resour. Stud. Unit. Univ. Hawaii Press, Honolulu.

Higashino, P.K., C.H. Lamoureux, R.L. Stemmermann, and F.R. Warshauer. 1977. Pohakuloa Training Area. In *A report on the botanical survey for the installation environmental impact statement*. U.S.

Army Support Command, Hawaii. Prepared for U.S. Corps Engineers by Environment Impact Study Corp., Honolulu.

Hillebrand, W.F. 1888. [Facsimile reprinted 1981.] *Flora of the Hawaiian Islands: a description of their phanerogams and vascular cryptogams*. Lubrecht & Cramer, Monticello, N.Y. 673 pp.

Hills, J. 1988. Saving Maui's ecosystem. *Maui, Inc*. 3(1):17-21, 59-62.

Hitchcock, A.S. 1922 [Reprinted 1974]. *The grasses of Hawaii*. Memoirs Bernice P. Bishop Mus., vol. 8, No. 3. Bishop Mus. Press, Honolulu, and Kraus Reprint Co., Millwood, New York. 131 pp. + plates.

Hodges, C.S. 1988. *Preliminary exploration for potential biological control agents for* Psidium cattleianum. Univ. Hawaii Coop. Natl. Park Resour. Stud. Unit Tech. Rept. 66. Botany Dept., Univ. Hawaii, Honolulu. 32 pp.

Hodges, C.S., Jr., and D.E. Gardner. 1985. *Myrica faya: potential biological control agents*. Univ. Hawaii Coop. Natl. Park Resour. Stud. Unit Tech. Rept. 54. Botany Dept., Univ. Hawaii, Honolulu. 37 pp.

Holden, C. 1985. Hawaiian rainforest being felled. *Science* 31:1073-1074.

Holt, R.A. 1989. Protection of natural habitats. pp. 168-174 In C.P. and D.B. Stone (eds.), *Conservation biology in Hawai'i*. Univ. Hawaii Coop. Natl. Park Resour. Stud. Unit. Univ. Hawaii Press, Honolulu.

Hommon, R.J. 1976. The formation of primitive states in pre-contact Hawaii. PhD Diss., Dept. Anthropology, Univ. Arizona. Hawaiian Collection, Mookini Library, Univ. Hawaii, Hilo. 356 pp.

Hommon, R.J. 1980. Historic resources of Kaho'olawe: the National Register multiple resources nomination overview. Unpubl. rept. Hawaiian Collection, Hamilton Library, Univ. Hawaii, Honolulu.

Horner, A. 1908. Letter to W.M. Giffard, Jan. 2, 1908. L.W. Bryan Forestry Collection, B.P. Bishop Museum, Honolulu. 5 pp.

Hosaka, E.Y. 1931. History of the Hawaiian forest. Unpubl. rept. Hawaii Collection, Mookini Library, Univ. Hawaii, Hilo. 25 pp.

Hosaka, E.Y. 1958. *Kikuyu grass in Hawaii*. Univ. Hawaii Agric. Extn. Svc. Circular 389. Coll. Agric., Univ. Hawaii, Honolulu. 18 pp.

Hosaka, E.Y., and J.C. Ripperton. 1944. *Legumes of the Hawaiian ranges*. Hawaii Agric. Expt. Sta. Bull. 93. Honolulu. 80 pp.

Hosaka, E.Y., and A. Thistle. 1954. *Noxious plants of the Hawaiian ranges*. Hawaii Agric. Expt. Sta. Bull. 62. Univ. Hawaii College Agric., Agric. Extension Svc. Honolulu. 39 pp.

Hostetler, H. 1970. Many island trees fall victims to imports. *The Sunday Star-Bulletin and Advertiser*, May 31, column: Looking ahead with Hostetler, p. A-10.

Hosmer, R.S. 1959. The beginning of five decades of forestry in Hawaii. *J. Forestry* 57(2):83-89.

Howarth, F.G. 1985. Impacts of alien land arthropods and mollusks on native plants and animals in Hawaii. pp. 149-179 In C.P. Stone and J.M. Scott (eds.), *Hawai'i's terrestrial ecosystems:*

preservation and management. Univ. Hawaii Coop. Natl. Park Resour. Stud. Unit. Univ. Hawaii Press, Honolulu.

Howarth, F.G., S.H. Sohmer, and W.D. Duckworth. 1988. Hawaiian natural history and conservation efforts. *Bioscience* 38(4):232-237.

Hugh, W.I., T. Tanaka, J.C. Nolan, Jr., and L.K. Fox. 1986. *The livestock industry in Hawaii.* Hawaii Inst. Trop. Agric. and Human Resour., Coll. Trop. Agric. and Human Resour., Univ. Hawaii. Information Text Series 025. Honolulu. 31 pp.

Hughes, F. 1989. The coupled effects of grass invasion and fire on the submontane seasonal ecosystem of Hawaii Volcanoes National Park. Senior Honors Thesis, Stanford Univ., Stanford, Calif. 29 pp. Files, Resour. Mgmt. Div., Hawaii Volcanoes Natl. Park.

Jacobi, J.D. In press a. A review of vegetation classification and mapping in Hawai'i. Part I. Literature Review. PhD Diss., Botany Dept., Univ. Hawaii, Honolulu.

Jacobi, J.D. In press b. Distribution and ecological relationships of the native upland vegetation on the island of Hawai'i. Part II. Original research. PhD Diss., Botany Dept., Univ. Hawaii, Honolulu.

Jacobi, J.D., and J.M. Scott. 1985. An assessment of the current status of native upland habitats and associated endangered species on the island of Hawai'i. pp. 3-22 In C.P. Stone and J.M. Scott (eds.), *Hawai'i's terrestrial ecosystems: preservation and management.* Univ. Hawaii Coop. Natl. Park Resour. Stud. Unit. Univ. Hawaii Press, Honolulu.

Jacobi, J.D., and F.R. Warshauer. 1975. A preliminary bioecological survey of the Ola'a Tract, Hawaii Volcanoes National Park. Prepared for Hawaii Natural History Assn. Unpubl. rept. Library, Hawaii Volcanoes Natl. Park. 73 pp. + appendices.

Jacobi, J.D., and F.R. Warshauer. In press. The current and potential distributions of six introduced plant species in upland habitats on the island of Hawaii. In C.P. Stone, C.W. Smith, and J.T. Tunison (eds.), *Alien plant invasions in native ecosystems of Hawai'i: management and research.* Univ. Hawaii Coop. Natl. Park Resour. Stud. Unit. Univ. Hawaii Press, Honolulu.

James, H.F., T.W. Stafford, Jr., D.W. Steadman, S.L. Olson, P.S. Martin, A.J.T. Jull, and P.C. McCoy. 1987. Radiocarbon dates on bones of extinct birds from Hawaii. *Proc. Natl. Acad. Sci. USA* 84:2350-2354.

Jenkins, I. 1983. *Hawaiian furniture and Hawaii's cabinet makers, 1820-1940.* The Daughters of Hawaii, Editions Ltd. Honolulu. 350 pp.

Judd, C.S. 1918. Forestry as applied in Hawaii. *Hawn. Forester and Agriculturist* 15(5):117-133.

Judd, C.S. 1924. Report on Honouliuli Forest Reserve. *Hawn. Forester and Agriculturist* 21:152-158.

Judd, C.S. 1927. The natural resources of the Hawaiian forest regions and their conservation. *Hawn. Forester and Agriculturist* 24:40-47.

Judd, C.S. 1931. Forestry in Hawaii for water conservation. *J. Forestry* 29(3):363-367.

Judd, C.S. 1932. Botanical discoveries. In Annual report 1931, Territorial Forester. *Hawn. Forester and Agriculturist* 29:15-16.

Juvik, J.O., and S.P. Juvik. 1984. Mauna Kea and the myth of multiple use, endangered species and mountain management in Hawaii. *Mountain Research and Devel.* 4(3):191-202.

Juvik, J.O., and L. Lawrence. 1982. Late Holocene vegetation history from Hawaiian peat deposits. p. 100 <u>In</u> C.W. Smith (ed.), *Proc. 4th Conf. Nat. Sci., Hawaii Volcanoes Natl. Park.* Botany Dept., Univ. Hawaii, Honolulu.

Juvik, J.O., and S.P. Juvik. In press. *Verbascum thapsus* L.: the spread and adaptation of a temperate weed in the montane tropics. <u>In</u> C.P. Stone, C.W. Smith, and J.T. Tunison (eds.), *Alien plant invasions in native ecosystems of Hawai'i: management and research.* Univ. Hawaii Coop. Natl. Park Resour. Stud. Unit. Univ. Hawaii Press, Honolulu.

Kami, H.T. 1966. Foods of rodents in the Hamakua District, Hawaii. *Pac. Sci.* 20(3):367-373.

Kaneshiro, K.Y., and C.R.B. Boake. 1987. Sexual selection and speciation: issues raised by Hawaiian *Drosophila. Trends in Ecol. and Evolution* 2(7):207-212.

Kartawinata, K., and D. Mueller-Dombois. 1972. Phytosociology and ecology of the natural dry-grass communities on Oahu, Hawaii. *Reinwardtia* 8(3):369-494.

Kelly, M. 1969. *Historical background of the South Point area, Ka'u, Hawai'i.* Pac. Anthropological Records 6. Anthropology Dept., B.P. Bishop Museum, Honolulu. 73 pp.

Kelly, M. 1983. *Na mala o Kona: gardens of Kona, a history of land use in Kona, Hawai'i.* Rept. 83-2. Anthropology Dept., B.P. Bishop Museum, Honolulu. 129 pp.

Kelly, M., and J.T. Clark. 1980. *Kawainui Marsh, O'ahu: historical and archaeological studies.* Rept. 80-3. Anthropology Dept., B.P. Bishop Museum, Honolulu. 82 pp.

Kelly, M., and S.N. Crozier. 1972. *Archaeological survey and excavations at Waiohinu drainage improvement project, Ka'u, island of Hawai'i.* Rept. 72-6. Prepared for Dept. Public Works, Co. Hawaii. Anthropology Dept., B.P. Bishop Museum, Honolulu. 56 pp.

Kirch, P.V. 1974. The chronology of early Hawaiian settlement. *Archeol. and Physical Anthro. in Oceania* 9(2):110-119.

Kirch, P.V. 1979. *Marine exploitation in prehistoric Hawai'i: archaeological excavations at Kalahuipua'a, Hawai'i Island.* Pac. Anthropological Records 29. Anthropology Dept., B.P. Bishop Museum, Honolulu. 235 pp.

Kirch, P.V. 1982. The impact of prehistoric Polynesians on the Hawaiian ecosystem. *Pac. Sci.* 36(1):1-14.

Kirch, P.V. 1983. Man's role in modifying tropical and subtropical Polynesian ecosystems. *Archaeol. Oceania* 18:26-31.

Kirch, P.V. 1985a. *Feathered gods and fishhooks. An introduction to Hawaiian archaeology and prehistory.* Univ. Hawaii Press, Honolulu. 349 pp.

Kirch, P.V. 1985b. Intensive agriculture in prehistoric Hawai'i: the wet and the dry. pp. 435-454 <u>In</u> I.S. Farrington (ed.), *Prehistoric intensive agriculture in the tropics, Part II.* British Archaeological Rept. Internatl. Series 232. Oxford, England.

Kirch, P.V., and M. Kelly (eds.). 1975. *Prehistory and ecology in a windward Hawaiian valley: Halawa Valley, Molokai.* Pac. Anthropological Records 24. Anthropology Dept., B.P. Bishop Museum, Honolulu. 207 pp.

Kivilaan, A., and R.S. Bandwrski. 1981. The one hundred-year period for Dr. Beal's seed viability experiment. *Amer. J. Bot.* 68(9):1290-1292.

Koebele, A. 1900. Hawaii's forest foes. *Thrum's Hawn. Almanac and Annual for 1901*, pp. 90-97.

Korte, K.H. 1963. Rodent damage in koa reproduction. Rept. State Forester, Hawaii Div., Kahului, Maui.

Kotzebue, O. von. 1821. *A voyage of discovery into the South Sea and Beering's Straits, for the purpose of exploring a northeast passage, undertaken in the years 1815-1818.* Translation edited by H.E. Lloyd. Vol. 3. London.

Krajina, V.J. 1963. Biogeoclimatic zones on the Hawaiian Islands. *Newsletter, Hawaiian Bot. Soc.* 2(7):93-98. [Reprinted in E.A. Kay (ed.), 1972. *A natural history of the Hawaiian Islands, selected readings,* pp. 273-277. Univ. Press Hawaii, Honolulu.]

Kramer, R.J. 1971. *Hawaiian land mammals.* Charles E. Tuttle Company, Tokyo, Japan. 347 pp.

Krauss, B.H. (compiler). n.d. *Ethnobotany of Hawaii.* Botany Dept., Univ. Hawaii, Honolulu. 246 pp.

Kuykendall, R.S. 1938. *The Hawaiian kingdom (1778-1854).* Vol. I. *Foundation and transformation.* Univ. Hawaii Press, Honolulu. 453 pp.

Kuykendall, R.S. 1953. *The Hawaiian kingdom (1854-1874).* Vol. II. *Twenty critical years.* Univ. Hawaii Press, Honolulu. 310 pp.

Kuykendall, R.S. 1967. *The Hawaiian kingdom (1874-1893).* Vol. III. *The Kalakaua dynasty.* Univ. Hawaii Press, Honolulu. 764 pp.

Lamoureux, C.H. 1967. The vascular plants of Kipahulu Valley, Maui. pp. 23-54 In R.E. Warner (ed.), *Scientific report of the Kipahulu Valley Expedition.* Sponsored by The Nature Conservancy.

LaRosa, A.M. 1984. *The biology and ecology of Passiflora mollissima in Hawaii.* Univ. Hawaii Coop. Natl. Park Resour. Stud. Unit Tech. Rept. 50. Botany Dept., Univ. Hawaii, Honolulu. 168 pp.

LaRosa, A.M. In press. The status of *Passiflora mollissima* (H.B.K.) Bailey in Hawai'i. In C.P. Stone, C.W. Smith, and J.T. Tunison (eds.), *Alien plant invasions in native ecosystems of Hawai'i: management and research.* Univ. Hawaii Coop. Natl. Park Resour. Stud. Unit. Univ. Hawaii Press, Honolulu.

LaRosa, A.M., C.W. Smith, and D.E. Gardner. 1985. Role of alien and native birds in the dissemination of firetree (*Myrica faya* Ait. -- Myricaceae) and associated plants in Hawaii. *Pac. Sci.* 39(4):372-378.

Leeper, J.R., and J.W. Beardsley, Jr. 1977. The biocontrol of *Psylla uncatoides* (Fessius and Klyver) (Homoptera: Psyllidae) on Hawaii. *Proc. Hawn. Entomol. Soc.* 22:307-321.

Lennox, C.G. 1948. Are forests essential to Hawaii's water economy? *Biennial Rept., Board of Commissioners of Agric. and Forestry.* Honolulu. 10 pp.

Lew, A.A. 1981. Land use and geothermal energy development in lower Puna, Hawaii. Dept. Research and Development, Hawaii Co., Hilo. Senior Thesis, Univ. Hawaii, Hilo. 65 pp.

Lewin, V., and G. Lewin. 1984. The Kalij pheasant, a newly established game bird on the island of Hawaii. *Wilson Bull.* 96(4):634-646.

Linney, G.K. 1986. *Coccinea grandis* (L.) Voigt: a new cucubitaceous weed in Hawai'i. *Newsletter, Hawaiian Bot. Soc.* 25(1):3-5.

Linney, G.K. 1989. An update on *Coccinea* (Cucurbitaceae) in Hawai'i. *Newsletter, Hawaiian Bot. Soc.* 28(2):34-35.

Little, E.L., Jr., and R.G. Skolmen. 1989. *Common forest trees of Hawaii (native and introduced).* Agric. Handbook 679. U.S. Forest Service, Washington, D.C. 321 pp.

Lockwood, J.P. 1985. 'Ohi'a chipping -- image, reality. [Viewpoint article]. *Hawaii Tribune-Herald*, Tues., June 18, p. 4.

Loope, L.L. In press. An overview of problems with introduced plant species in national parks and reserves of the United States. In C.P. Stone, C.W. Smith, and J.T. Tunison (eds.), *Alien plant invasions in native ecosystems of Hawai'i: management and research.* Univ. Hawaii Coop. Natl. Park Resour. Stud. Unit. Univ. Hawaii Press, Honolulu.

Loope, L.L., and C.F. Crivellone. 1986. *Status of the silversword in Haleakala National Park: past and present.* Univ. Hawaii Coop. Natl. Park Resour. Stud. Unit. Tech. Rept. 58. Botany Dept., Univ. Hawaii, Honolulu. 33 pp.

Loope, L.L., and D. Mueller-Dombois. 1989. Characteristics of invaded islands, with special reference to Hawai'i. pp. 257-280 In J.A. Drake, H.A. Mooney, F. DiCastri, R.H. Groves, FmJ. Kruger, M. Rejmànek, and M. Williamson (eds.)., *Biological invasions: a global perspective.* John Wiley and Sons, Chichester, U.K.

Loope, L.L., and P.G. Scowcroft. 1985. Vegetation response within exclosures in Hawaii: a review. pp. 377-402 In C.P. Stone and J.M. Scott (eds.), *Hawai'i's terrestrial ecosystems: preservation and management.* Univ. Hawaii Coop. Natl. Park Resour. Stud. Unit. Univ. Hawaii Press, Honolulu.

Loope, L.L., O. Hamann, and C.P. Stone. 1988. Comparative conservation biology of oceanic archipelagoes. *Bioscience* 38(4):272-282.

Loope, L.L., A.C. Medeiros, and B.H. Gagné. In press a. *Aspects of the history and biology of the montane bogs of Haleakala National Park.* Univ. Hawaii Coop. Natl. Park Resour. Stud. Unit Tech. Rept. Botany Dept., Univ. Hawaii, Honolulu.

Loope, L.L., R.J. Nagata, and A.C. Medeiros. In press b. Introduced plants in Haleakala National Park. In C.P. Stone, C.W. Smith, and J.T. Tunison (eds.), *Alien plant invasions in native ecosystems of Hawai'i: management and research.* Univ. Hawaii Coop. Natl. Park Resour. Stud. Unit. Univ. Hawaii Press, Honolulu.

Lyon, H.L. 1918. The forests of Hawaii. *Hawaii Planters' Record* 20:276-281.

Lyon, H.L. 1921. Letter to the Director, Experiment Station, Hawaii Sugar Planters' Association, Sept. 23. L.W. Bryan Forestry Collection, B.P. Bishop Museum, Honolulu. 3 pp.

Lyon, H.L. 1922. Letter to C.S. Judd, Executive Officer, Board of Commissioners of Agriculture and Forestry, March 17. L.W. Bryan Forestry Collection, B.P. Bishop Museum, Honolulu. 13 pp.

Lyon, H.L. 1923. Letter on forestry from Harold L. Lyon to H.P. Agee. pp. 48-55 In Hartt, C.E. (ed.). 1980. *Harold Lloyd Lyon, Hawaiian sugar botanist.* Harold L. Lyon Arboretum, Univ. Hawaii, Honolulu.

Lyon, H.L. 1930. The flora of Moanalua 100,000 years ago. *Proc. Hawn. Acad. Sci.* B.P. Bishop Museum Spec. Publ. 16:6-7.

Lyon, H.L. 1932. Letter to L.W. Bryan, May 17. L.W. Bryan Forestry Collection, B.P. Bishop Museum, Honolulu. 2 pp.

Lyon, H.L. 1946. Letter to L.W. Bryan, May 24. L.W. Bryan Forestry Collection, B.P. Bishop Museum, Honolulu. 1 p.

MacBride, L.R. 1975. *Practical folk medicine of Hawaii.* Petroglyph Press, Ltd., Hilo. 104 pp.

MacCaughey, V. 1917. Phytogeography of Manoa Valley. *Amer. J. Botany* 4:561-603.

MacCaughey, V. 1918. History of botanical exploration in Hawaii. *Hawn. Forester and Agriculturist* 15:388-396.

MacDonald, G.A., and A.T. Abbott. 1979. *Volcanoes in the sea; the geology of Hawaii.* Univ. Hawaii Press, Honolulu. 441 pp.

Maciolek, J.A., and R.E. Brock. 1974. *Aquatic survey of the Kona coast ponds, Hawaii Island.* Sea Grant Advisory Rept. AR-74-04. Univ. Hawaii Sea Grant Program, Honolulu. 20 pp. + appendices.

Macrae, J. 1972. *With Lord Byron at the Sandwich Islands in 1825. Being extracts from the M.S. diary of James Macrae, Scottish botanist.* Petroglyph Press, Hilo, Hawaii 87 pp.

Manning, A. 1986. Bishop and Dole on forest management. *'Elepaio* 46(15):163-165.

Markin, G.P., P.Y. Lai, and G.Y. Funasaki. In press. Status of biological control of weeds in Hawai'i and implications for managing native ecosystems. In C.P. Stone, C.W. Smith, and J.T. Tunison (eds.), *Alien plant invasions in native ecosystems of Hawai'i: management and research.* Univ. Hawaii Coop. Natl. Park Resour. Stud. Unit. Univ. Hawaii Press, Honolulu.

Markin, G.P., L.A. Dekker, J.A. Lapp, and R.F. Nagata. 1988. Distribution of gorse (*Ulex europaeus* L.), a noxious weed in Hawai'i. *Newsletter, Hawaiian Bot. Soc.* 27(3):110-117.

McEldowney, H. 1979. *Archeological and historical literature search and research design, lava flow control study, Hilo, Hawaii.* Prepared for U.S. Army Corps Engineer Div., Pacific Ocean. Anthropology Dept., B.P. Bishop Museum, Honolulu. 66 pp.

McEldowney, II. 1983. A description of major vegetation patterns in the Waimea-Kawaihae region during the early historic period. Rept. 16, pp. 407-448 In J.T. Clark and P.V. Kirch (eds.). 1983. *Archaeological investigations of the Mudlane-Waimea-Kawaihae Road corridor, island of Hawai'i. An interdisciplinary study of an environmental transect.* Rept. 83-1. Anthropology Dept., B.P. Bishop Museum, Honolulu.

McKeown, S. 1978. *Hawaiian reptiles and amphibians.* The Oriental Publ. Co., Honolulu. 80 pp.

Medeiros, A.C., L.L. Loope, and R. Hobdy. 1984. Remnant native vegetation at a lowland site near Kihei, Maui. pp. 78-82 In C.W. Smith (ed.), *Proc. 5th Conf. Nat. Sci., Hawaii Volcanoes Natl. Park.* Botany Dept., Univ. Hawaii, Honolulu.

Medeiros, A.C., Jr., L.L. Loope, and R.A. Holt. 1986. *Status of native flowering plant species on the south slope of Haleakala, East Maui, Hawaii.* Univ. Hawaii Coop. Natl. Park Resour. Stud. Unit Tech. Rept. 59. Botany Dept., Univ. Hawaii, Honolulu. 230 pp.

Medeiros, A.C., R.W. Hobdy, and L.L. Loope. 1989. Status of *Clidemia hirta* on Haleakala. *Newsletter, Hawaiian Bot. Soc.* 28(1):3-4.

Medeiros, A.C., R. Rydell, and L.L. Loope. 1988. Plant succession in native rain forest of East Maui following marijuana cultivation. *'Elepaio* 48(10):81-84.

Menzies, A. 1920. [W.F. Wilson, ed.] *Hawaii Nei 128 years ago.* The New Freedom, Honolulu. 199 pp.

Merrill, E.D. 1941. Man's influence on the vegetation of Polynesia, with special reference to introduced species. *Proc. 6th Pac. Sci. Congress, Calif.* 4:629-639.

Meyen, F.J.F. 1981. (Translated by A. Jackson. M.A. Pultz, ed.) *A botanist's visit to Oahu in 1831. Being the journal of Dr. F.J.F. Meyen's travels and observations about the island of Oahu.* Press Pacifica, Ltd., Kailua, Hawaii. 90 pp.

Morgan, J. 1983. *Hawaii, a geography.* Westview Press, Boulder, Colorado. 293 pp.

Morgan, J. 1989. Tourism. pp. 146-153 In C.P. Stone and D.B. Stone (eds.), *Conservation biology in Hawai'i.* Univ. Hawaii Coop. Natl. Park Resour. Stud. Unit. Univ. Hawaii Press, Honolulu.

Mueller-Dombois, D. 1966. The vegetation map and vegetation profiles. pp. 391-441 In M.S. Doty and D. Mueller-Dombois, *Atlas for bioecology studies in Hawaii Volcanoes National Park.* Hawaii Bot. Soc. Paper No. 2. Univ. Hawaii, Honolulu. 507 pp.

Mueller-Dombois, D. 1967. Ecological relations in the alpine and subalpine vegetation on Mauna Loa, Hawaii. *J. Indian Bot. Soc.* 46(4):403-411.

Mueller-Dombois, D. 1973. A non-adapted vegetation interferes with water removal in a tropical rain forest area in Hawaii. *Trop. Ecol.* 14(1):1-18.

Mueller-Dombois, D. 1981a. Fires in tropical ecosystems. pp. 137-176 In H.A. Mooney, T.M. Bonnicksen, N.L. Christensen, J.E. Lotan, and W.A. Reiners (eds.), *Fire regimes and ecosystem properties. Proc. Conf. Dec. 11-15, 1978, Honolulu, Hawaii.* U.S. Dept. Agriculture, Forest Service Gen. Tech. Rept. WO-26. Washington, D.C.

Mueller-Dombois, D. 1981b. Vegetation dynamics in a coastal grassland of Hawaii. *Vegetatio* 46:131-140.

Mueller-Dombois, D. 1985. The biological resource value of native forest in Hawaii with special reference to the tropical lowland rainforest at Kalapana. Rept. prepared for BioPower Corp. *'Elepaio* 45(10):95-101.

Mueller-Dombois, D. 1986. Perspectives for an etiology of stand-level dieback. *Ann. Rev. Ecol. Syst.* 17:221-243.

Mueller-Dombois, D. 1987. Forest dynamics in Hawaii. *Trends in Ecol. and Evolution* 2(7):216-220.

Mueller-Dombois, D., and V.J. Krajina. 1968. Comparison of east-flank vegetations on Mauna Loa and Mauna Kea, Hawaii. *Proc. Symp. Recent Adv. Trop. Ecol.* Part II, pp. 502-520.

Mueller-Dombois, D., and C.H. Lamoureux. 1967. Soil-vegetation relationships in Hawaiian kipukas. *Pac. Sci.* 21(2):286-299.

Mueller-Dombois, D., and G. Spatz. 1975. The influence of feral goats on the lowland vegetation in Hawaii Volcanoes National Park. *Phytocoenologica* 3(1):1-29.

Muench, M.N., R.K. Roberts, and F.S. Scott, Jr. 1984. *The economic viability of papaya farms in the Puna district.* Research Ser. 027, Hawaii Inst. Trop. Agric. and Human Resour., Coll. Trop. Agric. and Human Resour., Univ. Hawaii, Honolulu. 57 pp.

Mull, M. 1982. Convert a native forest to a eucalyptus farm? *'Elepaio* 43(1):4-5.

Mull, M. 1986a. No Na Leo 'Ole. *'Elepaio* 46(12):137-139.

Mull, M. 1986b. No Na Leo 'Ole; geothermal energy development on Big Island conservation lands. *'Elepaio* 46(18):193-194.

Mull, M. 1986c. No Na Leo 'Ole. *'Elepaio* 46(17):184-185.

Mull, M. 1986d. No Na Leo 'Ole. *'Elepaio* 46(8):85-86.

Mull, M. 1987. Conservation news. *'Elepaio* 47(8):81-83.

Munro, G.C. 1930. Molasses grass. *Hawn. Forester and Agriculturist* 27(2):61-64.

Murakami, G.M. 1983. Analysis of charcoal from archaeological contexts. Rept. 20, pp. 514-526 In J.T. Clark and P.V. Kirch (eds.). 1983. *Archaeological investigations of the Mudlane-Waimea-Kawaihae Road corridor, island of Hawai'i. An interdisciplinary study of an environmental transect.* Rept. 83-1. Anthropology Dept., B.P. Bishop Museum, Honolulu.

Myers, P. 1976. *Zoning Hawaii, an analysis of the passage and implementation of Hawaii's land classification law.* The Conservation Foundation, Washington, D.C. 110 pp.

Myhre, S.B. 1970. Kahoolawe. *Newsletter, Hawaiian Bot. Soc.* 9(4):21-26.

Nagata, K.M. 1985. Early plant introductions in Hawai'i. *Hawn. J. History.* 19:35-61.

Nakahara, L.M. 1980. Survey report on the yellow jackets, *Vespula pensylvanica* (Soussine) and *Vespula vulgaris* (L.) in Hawaii. Hawaii State Dept. Agric., Honolulu. Mimeo.

Nakahara, L.M., R.M. Burkhart, and G.Y. Funasaki. In press. Status and feasibility of *Clidemia* control [review and status of insects for biological control of *Clidemia hirta* in Hawai'i]. In C.P. Stone, C.W. Smith, and J.T. Tunison (eds.), *Alien plant invasions in native ecosystems of Hawai'i: management and research.* Univ. Hawaii Coop. Natl. Park Resour. Stud. Unit. Univ. Hawaii Press, Honolulu.

National Park Service. 1986a. Wildland fire management plan and environmental assessment, Hawaii Volcanoes National Park. An amendment to the natural resources management plan, 1985. Unpubl. rept. On file, Resour. Mgmt. Div., Hawaii Volcanoes Natl. Park. 61 pp.

National Park Service. 1986b. Natural Resources Management Program, Haleakala National Park. Unpubl. rept. On file, Research Div., Hawaii Volcanoes Natl. Park. pp. unnumbered.

National Park Service. 1989. Wildland fire management and environmental assessment, HAVO. An amendment to the natural resources management plan. Unpubl. rept. Resources Mgmt. Div., Hawaii Volcanoes Natl. Park.

Natural Area Reserves System Commission. 1989. Recommendation to establish the Kanaio Natural Area Reserve, Maui. Draft of August 3, 1989. 4 pp. + maps.

Neal, M.C. 1965. *In gardens of Hawaii*. B.P. Bishop Museum Special Publ. 50. Bishop Mus. Press, Honolulu. 924 pp.

Nelson, R.E. 1960. *Silk-oak in Hawaii . . . pest or potential timber?* Misc. Paper 47. U.S. Dept. Agric. Pacific Southwest Forest and Range Expt. Stn. U.S. Forest Service, Berkeley, Calif. 5 pp.

Nelson, R.E. 1963. Timber -- bigger than sugar? *Hawaii Business and Industry* 8(8):48-50.

Nelson, R.E. 1967. *Records and maps of forest types in Hawaii*. Resour. Bull. PSW-8, U.S. Dept. Agric. Pac. SW Forest and Range Expt. Sta. U.S. Forest Service, Berkeley, Calif. 22 pp.

Nelson, R.E., and E.M. Hornibrook. 1962. *Commercial uses and volume of Hawaiian tree fern*. Tech. Paper 73. Pacific Southwest Expt. Stn. U.S. Forest Service, Berkeley, Calif. 10 pp.

Nelson, R.E., and N. Honda. 1966. *Plantation timber on the island of Hawaii - 1965*. U.S. Forest Service Resour. Bull. PSW-3. Pac. SW Forest and Range Expt. Sta., Berkeley, Calif., and Hawaii State Dept. Land Nat. Resour., Div. Forestry, Honolulu. 52 pp.

Nelson, R.E., and P.R. Wheeler. 1963. *Forest resources of Hawaii -- 1961*. Hawaii State Dept. Land Nat. Resour., Div. Forestry, in cooperation with Pac. SW Forest and Range Expt. Sta., U.S. Forest Service, Honolulu. 48 pp.

Newman, A. 1984. Conservation update. *'Elepaio* 44(10):101-102.

Newman, T.S. 1969. Cultural adaptations to the island of Hawaii ecosystem: the theory behind the 1968 Lapakahi project. p. 3-14 In R. Pearson (ed.), *Archaeology on the island of Hawaii*. Asian and Pac. Archaeol. Ser. No. 3. Univ. Hawaii, Honolulu.

Newman, T.S. 1970. Fishing and farming on the island of Hawaii in A.D. 1778. Div. State Parks, Honolulu. Unpubl. rept. Hawaiian Collection, Mookini Library, Univ. Hawaii, Hilo. 305 pp.

Newman, T.S. 1972. Man in the prehistoric Hawaiian ecosystem. pp. 559-603 In E.A. Kay (ed.), *A natural history of the Hawaiian Islands, selected readings*. Univ. Press Hawaii, Honolulu.

Norbeck, E. 1959. *Pineapple town, Hawaii*. Univ. Calif. Press, Berkeley, Calif. 159 pp.

Obata, J.K. 1985a. Another noxious melastome? *Oxyspora paniculata*. *Newsletter, Hawaiian Bot. Soc.* 24:25-26.

Obata, J.K. 1985b. The declining forest cover of the Ko'olau summit. *Newsletter, Hawaiian Bot. Soc.* 24:41-42.

Obata, J.K. 1986. The demise of a species: *Urera kaalae*. *Newsletter, Hawaiian Bot. Soc.* 25(2):74-75.

Olson, S.L., and H.F. James. 1982. Prodromus of the fossil avifauna of the Hawaiian Islands. *Smithsonian Contrib. Zool.* 365:1-59.

Parman, T.T., and K. Wampler. 1977. *The Hilina Pali fire: a controlled burn exercise.* Univ. Hawaii Coop. Natl. Park Resour. Stud. Unit Tech. Rept. 18. Botany Dept., Univ. Hawaii, Honolulu. 28 pp.

Pemberton, C.E. 1964. Highlights in the history of entomology in Hawaii, 1778-1963. *Pac. Insects* 6:689-729.

Perkins, R.C.L. 1903. Vertebrata. pp. 365-466 In D. Sharp (ed.), *Fauna Hawaiiensis, or the zoology of the Sandwich (Hawaiian) Isles.* Vol. I, pt. IV. Univ. Press, Cambridge, England.

Perkins, R.C.L. 1913. Introduction. pp. xv-ccxxviii In D. Sharp (ed.), *Fauna Hawaiiensis, or the zoology of the Sandwich (Hawaiian) Isles.* Vol. 1. Univ. Press, Cambridge, England.

Perkins, R.C.L., and O.H. Swezy. 1924. The introduction into Hawaii of insects that attack lantana. *Entomol. Ser. Bull. Expt. Stn. Hawaii Sugar Planters Assoc.* 16.

Philipp, P.F. 1953. *Diversified agriculture of Hawaii.* Univ. Hawaii Press, Honolulu. 226 pp.

Plasch, B.S. 1981. *Hawaii's sugar industry: problems, outlook, and urban growth issues.* Hawaii State Dept. Plan. Econ. Dev., Honolulu. 254 pp.

Plucknett, D.L., and B.C. Stone. 1961. The principal weedy Melastomaceae in Hawaii. *Pac. Sci.* 15(2):301-303.

Pope, W.T. 1926. *Banana culture in Hawaii.* Hawaii Agric. Expt. Sta. Bull. No. 55. Honolulu. U.S. Govt. Printing Office, Washington, D.C. 48 pp.

Porter, J.R. 1972. *Hawaiian names for vascular plants.* Hawaii Agric. Expt. Sta., Coll. Trop. Agric. Dept. Paper 1. 64 pp.

Powell, L. 1984. The Mauna Kea silversword: a species on the brink of extinction. *Newsletter, Hawaiian Bot. Soc.* 24:44-57.

Powell, L. 1985. Hapu'u farms and the Bishop Estate. *'Elepaio* 45(9):87.

Powell, L., and R. Warshauer. 1985a. No Na Leo 'Ole. *'Elepaio* 45(12):131-135.

Powell, L., and R. Warshauer. 1985b. No Na Leo 'Ole. *'Elepaio* 45(10):103-105.

Pratt, H.D., P.L. Bruner, and D.G. Berrett. 1987. *The birds of Hawaii and the tropical Pacific.* Princeton Univ. Press, Princeton, New Jersey. 409 pp. + 45 plates.

Pukui, M.K., and S.H. Elbert. 1981. *Hawaiian dictionary. Hawaiian-English, English-Hawaiian.* Univ. Hawaii Press, Honolulu. 402 + 188 pp.

Pukui, M.K., S.H. Elbert, and E.T. Mookini. 1986. *Place names of Hawaii,* 2nd ed. Univ. Hawaii Press, Honolulu. 289 pp.

Pyne, S.J. 1982. *Fire in America. A cultural history of wildland and rural fire.* Princeton Univ. Press, New Jersey. 654 pp.

Ramsay, G.W. 1978. A review of the effect of rodents on the New Zealand invertebrate fauna. pp. 89-95 In P.R. Dingwall, I.A.E. Atkinson, and C. Hay (eds.), *The ecology and control of rodents in New Zealand native reserves*. New Zealand Dept. Lands Surv. Info. Ser. 4.

Reimer, N.J. 1985. An evaluation of the status and effectiveness of *Liothrips urichi* Karny (Thysanoptera: Phlaeothripidae) and *Blepharomastix ebulealis* Guenee (Lepidoptera: Pyralidae) on *Clidemia hirta* (L.) D. Don in Oʻahu forests. PhD Diss., Univ. Hawaii, Honolulu. 159 pp.

Richmond, T. de A., and D. Mueller-Dombois. 1972. Coastline ecosystems on Oahu, Hawaii. *Vegetatio* 25:367-400.

Ripperton, J.C., and E.Y. Hosaka. 1942. *Vegetation zones of Hawaii*. Hawaii Agric. Expt. Stn. Bull. No. 89, Honolulu. 60 pp.

Robyns, W., and S.H. Lamb. 1939. Preliminary ecological survey of the island of Hawaii. *Bull. Jardin Botan. de l'Etat a Bruxelles* 15(3):241-293.

Rock, J.F. 1913 [Reprinted 1974, with annotations.]. *The indigenous trees of the Hawaiian Islands*. Pac. Trop. Bot. Gdn., Lawai, Kauai, Hawaii, and Charles E. Tuttle Co., Rutland, Vermont and Tokyo, Japan. 548 pp.

Rosendahl, P.H. 1972. Aboriginal agriculture and residence patterns in upland Lapakahi, island of Hawaii. PhD Diss., Univ. Hawaii, Honolulu. 558 pp.

Rosendahl, P.H. 1973. *Archaeological salvage of the Ke-ahole to Anaehoomalu section of the Kailua-Kawaihae Road (Queen Kaahumanu Highway), island of Hawaii*. Rept. 73-2. Hawaii Historic Preservation, Anthropology Dept., B.P. Bishop Museum, Honolulu. 131 pp.

Rosendahl, P.H. 1974. The Hawaiian agricultural system at Lapakahi, Hawaii Island. Paper presented at 94th Annual Conf. Hawaii Sugar Planters' Assn., Mauna Kea Beach Hotel. 30 pp. Hawaiian Collection, Mookini Library, Univ. Hawaii, Hilo.

Rosendahl, P.H. (ed.). 1976. *Archaeological investigations in upland Kaneohe: survey and salvage excavations in the upper Kamoʻoaliʻi stream drainage area, Kaneohe, Koʻolaupoko, Oʻahu, Hawaiʻi*. Rept. Ser. 76-1. Anthropology Dept., B.P. Bishop Museum, Honolulu. 250 pp.

Rotondo, G.M., V.G. Springer, G.A.J. Scott, and S.O. Schlanger. 1981. Plate movement and island integration -- a possible mechanism in the formation of endemic biotas, with special reference to the Hawaiian Islands. *Systematic Zool.* 30(1):12-21.

Rundel, P.W. 1980. The ecological distribution of C_4 and C_3 grasses in the Hawaiian Islands. *Oecologia* 45:354-359.

Russell, C.A. 1980. Food habits of the roof rat *(Rattus rattus)* in two areas of Hawaii Volcanoes National Park. pp. 269-272 In C.W. Smith (ed.), *Proc. 3rd Conf. Nat. Sci., Hawaii Volcanoes Natl. Park*. Botany Dept., Univ. Hawaii, Honolulu.

St. John, H. 1947. The history, present distribution, and abundance of sandalwood on Oahu, Hawaiian Islands. Hawaiian Plant Studies 14. *Pac. Sci.* 1:5-20.

St. John, H. 1959. Botanical novelties on the island of Niihau, Hawaiian Islands. Hawaiian Plant Studies 25. *Pac. Sci.* 13:156-190.

St. John, H. 1973. *List and summary of the flowering plants in the Hawaiian Islands.* Pac. Trop. Bot. Gdn. Mem. No. 1. Lawai, Kauai, Hawaii. 519 pp.

St. John, H. 1976. New species of Hawaiian plants collected by David Nelson in 1779. Hawaiian Plant Studies 52. *Pac. Sci.* 30(1):7-44.

St. John, H. 1978. The first collection of Hawaiian plants by David Nelson in 1779. Hawaiian Plant Studies 55. *Pac. Sci.* 32(3):315-324.

St. John, H. 1979. The vegetation of Hawaii as seen on Captain Cook's voyage in 1779. *Pac. Sci.* 33(1):79-83.

St. John, H. 1982. Vernacular plant names used on Niʻihau Island. Hawaiian Plant Studies 69. *Occas. Papers Bernice P. Bishop Museum* 25(3):1-10.

St. John, H., and M. Titcomb. 1983. The vegetation of the Sandwich Islands as seen by Charles Gaudichaud in 1819, a translation with notes of Gaudichaud's "Iles Sandwich." *Occas. Papers Bernice P. Bishop Mus.* 25(9):1-16.

Santos, G.L., L.W. Cuddihy, and C.P. Stone. 1989a. Banana poka control in Hawaii Volcanoes National Park. p. 137 In *Progress Rept., Western Soc. Weed Sci. Conf., Honolulu, Hawaii March 14-16, 1989.*

Santos, G.L., L.W. Cuddihy, and C.P. Stone. 1989b. Cut stump, frill, and basal bark treatments of triclopyr on strawberry guava. p. 134 In *Progress Rept., Western Soc. Weed Sci. Conf., Honolulu, Hawaii March 14-16, 1989.*

Santos, G.L., L.W. Cuddihy, and C.P. Stone. 1989c. Cut-stump and frill treatments of firetree in Hawaii Volcanoes National Park. pp. 135-136 In *Progress Rept., Western Soc. Weed Sci. Conf., Honolulu, Hawaii March 14-16, 1989.*

Santos, G.L., D. Kageler, D.E. Gardner, and C.P. Stone. 1986. *Herbicidal control of selected alien plant species in Hawaii Volcanoes National Park: a preliminary report.* Univ. Hawaii Coop. Natl. Park Resour. Stud. Unit Tech. Rept. 60. Botany Dept., Univ. Hawaii. 54 pp.

Schmitt, R.C. 1968. *Demographic statistics of Hawaii: 1778-1965.* Univ. Hawaii Press, Honolulu. 271 pp.

Schmitt, R.C. 1971. New estimates of the pre-censal population of Hawaii. *J. Polynesian Soc.* 80:237-243.

Schmitt, R.C. 1977. *Historical statistics of Hawaii.* Univ. Press Hawaii, Honolulu. 679 pp.

Schmitt, R.C. 1989. Comment. pp. 114-121 In D.E. Stannard, *Before the horror: the population of Hawaiʻi on the eve of western contact.* Social Science Research Institute, Univ. Hawaii, Honolulu.

Schubert, T.H., and C.D. Whitesell. 1985. *Species trials for biomass plantations in Hawaii: a first appraisal.* Research Paper PSW-176, Pac. SW Forest and Range Expt. Sta. U.S. Forest Service, Berkeley, Calif. 13 pp.

Scott, F.S., Jr., and H.K. Marutani. 1982. *Economic viability of small macadamia nut farms in Kona.* Research Ser. 009, Hawaii Instit. Trop. Agric. Human Resour., Coll. Trop. Agric. Human Resour., Univ. Hawaii, Honolulu. 34 pp.

Scott, J.M., S. Mountainspring, F.L. Ramsey, and C.B. Kepeler. 1986. *Forest bird communities of the Hawaiian Islands: their dynamics, ecology, and conservation*. Studies in Avian Biol. No. 9. Cooper Ornithol. Soc., Calif. 431 pp.

Scowcroft, P.G. 1983. Tree cover changes in mamane *(Sophora chrysophylla)* forests grazed by sheep and cattle. *Pac. Sci.* 37(2):109-119.

Scowcroft, P.G., and J.G. Giffin. 1983. Feral herbivores suppress mamane and other browse species on Mauna Kea, Hawaii. *J. Range Manage.* 36(5):638-645.

Scowcroft, P.G., and R. Hobdy. 1986. Recovery of montane koa parkland vegetation protected from feral goats. *Biotropica* 19:208-215.

Scowcroft, P.G., and R.E. Nelson. 1976. *Disturbance during logging stimulates regeneration of koa*. USDA Forest Service Research Note PSW-306. Pac. SW Forest and Range Expt. Sta., Berkeley, Calif. 7 pp.

Scowcroft, P.G., and H.F. Sakai. 1983. Impact of feral herbivores on mamane forests of Mauna Kea, Hawaii: bark stripping and diameter class structure. *J. Range Manage.* 36(4):495-498.

Scowcroft, P.G., and H.F. Sakai. 1984. Stripping of *Acacia koa* bark by rats on Hawaii and Maui. *Pac. Sci.* 38(1):80-86.

Selling, O.H. 1948. *Studies in Hawaiian pollen statistics, part III. On the late quarternary history of the Hawaiian vegetation*. B.P. Bishop Museum Spec. Publ. 39. Bishop Museum Press, Honolulu. 154 pp.

Simon, C. 1987. Hawaiian evolutionary biology: an introduction. *Trends in Ecol. and Evolution* 2(7):175-178.

Skolmen, R.G. 1974. *Some woods of Hawaii -- properties and uses of 16 commercial species*. Gen. Tech. Rept. PSW-8. Pac. SW Forest and Range Expt. Sta., U.S. Forest Service, Berkeley, Calif. 30 pp.

Skolmen, R.G. 1979. *Plantings on the forest reserves of Hawaii 1910-1960*. Instit. Pacific Islands Forestry, Pac. SW Forest and Range Expt. Sta., U.S. Forest Service, Honolulu. 441 pp.

Skolmen, R.G. 1986a. Where koa can be grown. Paper presented at Koa Forest Conference, Hilo, Hawaii, December 17-19, 1986.

Skolmen, R.G. 1986b. *Performance of Australian provenances of Eucalyptus grandis and Eucalyptus saligna in Hawaii*. U.S. Forest Service Research Paper PSW-181. Pacific Soutwest Forest and Range Expt. Stn., U.S. Forest Service, Berkeley, Calif. 8 pp.

Skolmen, R.G., and D.M. Fujii. 1980. Growth and development of a pure stand of koa *(Acacia koa)* at Keauhou-Kilauea. pp. 301-310 In C.W. Smith (ed.), *Proc. 3rd Conf. Nat. Sci., Hawaii Volcanoes Natl. Park*. Botany Dept., Univ. Hawaii, Honolulu.

Skottsberg, C. 1941. The flora of the Hawaiian Islands and the history of the Pacific Basin. *Proc. 6th Pac. Sci. Congress* 4:685-707.

Skottsberg, C. 1953. Report on the standing committee for the protection of nature in and around the Pacific for the years 1939-1948. pp. 586-597 In *Proc., 7th Pac. Sci. Cong.*, Auckland, New Zealand.

Smathers, G.A. 1967. A preliminary survey of the phytogeography of Kipahulu Valley. pp. 55-86 In R.E. Warner (ed.), *Scientific Report of the Kipahulu Valley Expedition, Maui, Hawaii, 2 August - 31 August, 1967*. Sponsored by The Nature Conservancy.

Smathers, G.A., and D.E. Gardner. 1979. Stand analysis of an invading firetree (*Myrica faya* Aiton) population, Hawaii. *Pac. Sci.* 33(3):239-355.

Smith, C.W. 1985. Impacts of alien plants on Hawai'i's native biota. pp. 180-250 In C.P. Stone and J.M. Scott (eds.), *Hawai'i's terrestrial ecosystems: preservation and management*. Univ. Hawaii Coop. Natl. Park Resour. Stud. Unit. Univ. Hawaii Press, Honolulu.

Smith, C.W. 1988. Notes on weeds in Hawaii. *Newsletter, Hawaiian Bot. Soc.* 27(4):130-131.

Smith, C.W. 1989a. Controlling the flow of non-native species. pp. 139-145 In C.P. Stone and D.B. Stone (eds.), *Conservation biology in Hawai'i*. Univ. Hawaii Coop. Natl. Park Resour. Stud. Unit. Univ. Hawaii Press, Honolulu.

Smith, C.W. 1989b. Non-native plants. pp. 60-69 In C.P. Stone and D.B. Stone (eds.), *Conservation biology in Hawai'i*. Univ. Hawaii Coop. Natl. Park Resour. Stud. Unit. Univ. Hawaii Press, Honolulu.

Smith, C.W. In press. Distribution, status, phenology, rate of spread, and management of *Clidemia hirta* in Hawai'i. In C.P. Stone, C.W. Smith, and J.T. Tunison (eds.), *Alien plant invasions in native ecosystems of Hawai'i: management and research*. Univ. Hawaii Coop. Natl. Park Resour. Stud. Unit. Univ. Hawaii Press, Honolulu.

Smith, C.W., and J.T. Tunison. In press. Fire and alien plants in Hawai'i: research and management implications for native ecosystems. In C.P. Stone, C.W. Smith, and J.T. Tunison (eds.), *Alien plant invasions in native ecosystems of Hawai'i: management and research*. Univ. Hawaii Coop. Natl. Park Resour. Stud. Unit. Univ. Hawaii Press, Honolulu.

Smith, C.W., T.T. Parman, and K. Wampler. 1980. Impact of fire in a tropical submontane seasonal forest. pp. 313-324 In *Proc. 2nd Conf. Sci. Research in Natl. Parks*. vol. 10, Fire Ecology, San Francisco, Calif., Nov. 26-30, 1979.

Soehren, L.J., and T.S. Newman. 1968. *The archaeology of Kealakekua*. Special Publ., Dept. Anthropology, B.P. Bishop Museum, and Univ. Hawaii. Honolulu. Figure reproduced in P.V. Kirch, 1985.

Sohmer, S.H. 1976. Kalua'a Gulch revisited. *Newsletter, Hawaiian Bot. Soc.* 15(1):23-24.

Sorenson, J.C. 1977. *Andropogon virginicus* (broomsedge). *Newsletter, Hawaiian Bot. Soc.* 16:7-22.

Spatz, G. 1973. A comparison of the efficiency of pasture farming and koa forestry on the Keauhou Ranch on Hawaii. Institut für Grünlandlehre der Technischen Universität, München. Unpubl. ms. in files of D. Mueller-Dombois, Botany Dept., Univ. Hawaii, Honolulu.

Spatz, G., and D. Mueller-Dombois. 1973. The influence of feral goats on koa tree reproduction in Hawaii Volcanoes National Park. *Ecology* 54(4):870-876.

Spence, G.E., and S.L. Montgomery. 1976. Ecology of the dryland forest at Kanepu'u, island of Lanai. *Newsletter, Hawaiian Bot. Soc.* 15:62-80.

Spriggs, M. 1985. Prehistoric man-induced landscape enhancement in the Pacific: examples and implications. pp. 409-432 In I.S. Farrington (ed.), *Prehistoric intensive agriculture in the tropics.* Part I. Bar Internatl. Ser. 232, Oxford, England.

Stannard, D.E. 1989. *Before the horror: the population of Hawai'i on the eve of western contact.* Social Science Research Institute, Univ. Hawaii, Honolulu. 149 pp.

Stemmermann, M. 1983. Kahauale'a geothermal project. *'Elepaio* 43(8):63-65.

Stewart, O.C. 1962. Fire as the first great force employed by man. pp. 115-133 In W.L. Thomas, Jr. (ed.), *Man's role in changing the face of the earth.* Univ. Chicago Press, Chicago, Illinois.

Stone, B.C. 1959. Natural and cultural history report on the Kalapana Extension of the Hawaii National Park. Vol. II. Natural History Report, Botany. B.P. Bishop Museum, Honolulu. 67 pp. Mimeo.

Stone, C.P. 1985. Alien animals in Hawai'i's native ecosystems: toward controlling the adverse effects of introduced vertebrates. pp. 251-297 In C.P. Stone and J.M. Scott (eds.), *Hawai'i's terrestrial ecosystems: preservation and management.* Univ. Hawaii Coop. Natl. Park Resour. Stud. Unit. Univ. Hawaii Press, Honolulu.

Stone, C.P. In press. Feral pig *(Sus scrofa)* research and management in Hawai'i. In F. Spitz (ed.), *Biologie des suides.* Inst. Nat. Recherche Agronomique, France. [Presented IV Theriological Congress, Edmonton, Alberta, Canada, Aug. 1985.]

Stone, C.P., and L.L. Loope. 1987. Reducing negative effects of introduced animals on native biotas in Hawaii: what is being done, what needs doing, and the role of national parks. *Environ. Conserv.* 14(3):245-258.

Stone, C.P., and D.D. Taylor. 1984. Status of feral pig management and research in Hawaii Volcanoes National Park. pp. 106-111 In C.W. Smith (ed.), *Proc. 5th Conf. Nat. Sci., Hawaii Volcanoes Natl. Park.* Botany Dept., Univ. Hawaii, Honolulu.

Stone, C.P., P.C. Banko, P.K. Higashino, and F.G. Howarth. 1984. Interrelationships of alien and native plants and animals in Kipahulu Valley, Haleakala National Park: a preliminary report. pp. 91-105 In C.W. Smith (ed.), *Proc. 5th Conf. Nat. Sci., Hawaii Volcanoes Natl. Park.* Botany Dept., Univ. Hawaii, Honolulu.

Stone, C.P., P.K. Higashino, L.W. Cuddihy, and S.J. Anderson. In press. *Preliminary survey of feral ungulate and alien and rare plant occurrence on Hakalau Forest National Wildlife Refuge.* Univ. Hawaii Coop. Natl. Park Resour. Stud. Unit Tech. Rept. Botany Dept., Univ. Hawaii, Honolulu.

Street, J.M. 1989. Soils in Hawai'i. pp. 17-23 In C.P. Stone and D.B. Stone (eds.), *Conservation biology in Hawai'i.* Univ. Hawaii Coop. Natl. Park Resour. Stud. Unit. Univ. Hawaii Press, Honolulu.

Tagawa, T. 1976. Endangered species in Hawaii, effect on other resource management. *Newsletter, Hawaiian Bot. Soc.* 15(1):7-14.

Taylor, D., and L. Katahira. 1988. Radio telemetry as an aid in eradicating remnant feral goats. *Wildl. Soc. Bull.* 16(3):297-299.

Telfer, T.C. 1982. Status, trends and utilization of game mammals and their associated habitats on the island of Kauai. Hawaii State Dept. Land Nat. Resour., Div. Forestry and Wildlife, Proj. W-17-R-17. Mimeo.

Telfer, T.C. 1988. Status of the black-tailed deer on Kauai. *1988 Trans. Western Sect. Wildlife Soc.* 24:53-60.

TenBruggencate, J. 1986a. Plants once covered Kahoʻolawe's scars. *The Honolulu Advertiser*, Thurs., December 23, pp. B-1, B-2.

TenBruggencate, J. 1986b. Koster's curse to get a curse of its own. *The Honolulu Advertiser*, Fri., Oct. 10, p. A-12.

TenBruggencate, J. 1988a. Private sandalwood logging has state upset. *The Honolulu Advertiser*, Thurs., Sept. 27.

TenBruggencate, J. 1988b. Salvaging Mauna Loa's historic sandalwood crop. *The Honolulu Star-Bulletin and Advertiser*, Sun., Oct. 2, pp. A-3, A-6.

The Nature Conservancy of Hawaii. 1987. *Biological overview of Hawaii's Natural Area Reserves System.* Hawaii Heritage Program. Prepared for Hawaii State Dept. Land Nat. Resour. Honolulu. 40 pp. + appendices.

Thomas, D.M. 1985. *Geothermal resources assessment in Hawaii. Assessment of geothermal resources in Hawaii: No. 7.* Final rept. Prepared for Western States Coop. Direct Heat Resour. Assessment. Hawaii Inst. Geophysics, Univ. Hawaii, Honolulu. 115 pp.

Tisdell, C.A. 1982. *Wild pigs: environmental pest or economic resource.* Pergamon Press, Sydney, Australia. 445 pp.

Tomich, P.Q. 1986. *Mammals in Hawaiʻi: a synopsis and notational bibliography,* 2nd ed. B.P. Bishop Museum Special Publ. 76. Bishop Museum Press, Honolulu. 375 pp.

Towill, R.M., Corporation. 1982. *Environmental Impact Statement for the Kahaualeʻa Geothermal Project, district of Puna, island of Hawaii, state of Hawaii.* Prepared for True/Mid-Pacific Geothermal Venture and Trustees, Estate of James Campbell. Honolulu. 240 pp. + appendices.

Tuggle, H.D., R. Cordy, and M. Child. 1978. Volcanic glass hydration -- rind age determination for Bellows Dune, Hawaii. *New Zealand Archaeol. Assn. Newsletter* 21:37-77.

Tunison, J.T. In press a. Strategies and successes in controlling alien plants in an Hawaiian National Park. In *Proc. Symposium on Exotic Pest Plants, Miami, Florida, November 1988.*

Tunison, J.T. In press b. Fountain grass *(Pennisetum setaceum)* control in Hawaiʻi Volcanoes National Park: effort, economics, and feasibility. In C.P. Stone, C.W. Smith, and J.T. Tunison (eds.), *Alien plant invasions in native ecosystems of Hawaiʻi: management and research.* Univ. Hawaii Coop. Natl. Park Resour. Stud. Unit. Univ. Hawaii Press, Honolulu.

Tunison, T., M. Gates, and C. Zimmer. 1989. The fountain grass control program in Hawaii Volcanoes National Park: a progress report. Unpubl. rept. Resour. Mgmt. Div., Hawaii Volcanoes Natl. Park. 17 pp.

Tunison, T., and J. Leialoha. 1988. The spread of fire in alien grasses after lightning strikes in Hawaii Volcanoes National Park. *Newsletter, Hawaiian Bot. Soc.* 27(3):102-109.

Tunison, J.T., G.L. Behnke, and T.S.L. Lau. In press. *Plant succession after eleven years in the Namakani Paio burn.* Univ. Hawaii Coop. Natl. Park Resour. Stud. Unit Tech. Rept. Botany Dept., Univ. Hawaii, Honolulu.

135

Tunison, J.T., L.D. Whiteaker, L.W. Cuddihy, A.M. LaRosa, A.D. Kageler, M.R. Gates, N.G. Zimmer, and L. Stemmermann. In press. *The distribution of selected localized alien plants in Hawaii Volcanoes National Park.* Univ. Hawaii Coop. Natl. Park Resour. Stud. Unit Tech. Rept. Botany Dept., Univ. Hawaii, Honolulu.

Twibell, J. 1973. The ecology of rodents in the Tonga Islands. *Pac. Sci.* 27(1):92-98.

U.S. Fish and Wildlife Service. 1987. *Endangered and threatened wildlife and plants.* 50 CFR 17.11 and 17.12. U.S. Govt. Printing Office, Washington, D.C. 31 pp.

Uyehara, M. 1977. *The Hawaii ceded land trusts, their use and misuse.* Hawaiiana Almanac Publ. Co., Honolulu. 65 pp.

Vandercook, J.W. 1939. *King Cane, the story of sugar in Hawaii.* Harper and Brothers, Publ., New York and London. 192 pp.

van Riper, C., III. 1980. The phenology of the dryland forest of Mauna Kea, Hawaii, and the impact of recent environmental perturbations. *Biotropica* 12(4):282-291.

van Riper, S.G., and C. van Riper III. 1982. *A field guide to the mammals of Hawaii.* Oriental Publ. Co., Honolulu.

Vitousek, P.M. In press. Effects of alien plants on native ecosystems. In C.P. Stone, C.W. Smith, and J.T. Tunison (eds.), *Alien plant invasions in native ecosystems of Hawai'i: management and research.* Univ. Hawaii Coop. Natl. Park Resour. Stud. Unit. Univ. Hawaii Press, Honolulu.

Vitousek, P.M., L.L. Loope, and C.P. Stone. 1987. Introduced species in Hawaii: biological effects and opportunities for ecological research. *Trends in Ecol. and Evolution* 2(7):224-227.

Vitousek, P.M., L.R. Walker, L.D. Whiteaker, D. Mueller-Dombois, and P.A. Matson. 1987. Biological invasion by *Myrica faya* alters ecosystem development in Hawaii. *Science* 238:802-804.

Vogl, R.J. 1969. The role of fire in the evolution of the Hawaiian flora and vegetation. pp. 5-60 In *Proc. 9th Annual Tall Timbers Fire Ecol. Conf.*, Tallahassee, Fla.

Vogl, R.J., and J. Henrickson. 1971. Vegetation of an alpine bog on East Maui, Hawaii. *Pac. Sci.* 25(4):475-483.

Wagner, W.L., D.R. Herbst, and S.H. Sohmer. In press. *Manual of the flowering plants of Hawai'i.* Bishop Museum and Univ. Hawaii Presses, Honolulu.

Wagner, W.L., D.R. Herbst, and R.S.N. Yee. 1985. Status of the native flowering plants of the Hawaiian Islands. pp. 23-74 In C.P. Stone and J.M. Scott (eds.), *Hawai'i's terrestrial ecosystems: preservation and management.* Univ. Hawaii Coop. Natl. Park Resour. Stud. Unit. Univ. Hawaii Press, Honolulu.

Ward, D. 1988. The tragic logging of isle sandalwood trees. [Viewpoint article] *Hawaii Tribune-Herald*, Sun., Sept. 15.

Warner, R.E. 1960. A forest dies on Mauna Kea. *Pac. Discovery* 13:6-14.

Warshauer, F.R. 1974. Biological survey of Kealakomo and vicinity affected by 1969-1973 lava-generated wildfires, Hawaii Volcanoes National Park. Unpubl. rept. Library, Hawaii Volcanoes Natl. Park, and Resour. Mgmt. Div., Hawaii Volcanoes Natl. Park. 41 pp.

Warshauer, F.R. 1977. The Kalapana extension of Hawaii Volcanoes National Park: its variety, vegetation, and value. *Newsletter, Hawaiian Bot. Soc.* 27(3):102-109.

Warshauer, F.R. 1984. Cash in your chips, or where have all the forests gone? *'Elepaio* 45(6):48-51.

Warshauer, F.R. 1986. Forest reserve legislation. *'Elepaio* 46(7):77-79.

Warshauer, F.R. 1988. Analysis: Watersheds and forest reserves. Unpubl. rept. submitted to State Representative Virginia Isbell. 6 pp.

Warshauer, F.R., and J.D. Jacobi. 1982. Distribution and status of *Vicia menziesii* (Leguminosae), Hawai'i's first officially listed endangered plant species. *Biol. Conserv.* 23:111-126.

Warshauer, F.R., J.D. Jacobi, A.M. LaRosa, J.M. Scott, and C.W. Smith. 1983. *The distribution, impact, and potential management of the introduced vine Passiflora mollissima (Passifloraceae) in Hawai'i.* Univ. Hawaii Coop. Natl. Park Resour. Stud. Unit Tech. Rept. 48. Botany Dept., Univ. Hawaii, Honolulu. 39 pp.

Weisler, M., and P.V. Kirch. 1982. *The archaeological resources of Kawela, Molokai: their nature, significance and management.* Anthropology Dept., B.P. Bishop Museum, Honolulu. 100 pp. + appendices.

Welch, D.J. 1983. Archaeological investigations in section 2. Report 5, pp. 138-180 In J.T. Clark and P.V. Kirch (eds.). 1983. *Archaeological investigations of the Mudlane-Waimea-Kawaihae Road corridor, island of Hawai'i. An interdisciplinary study of an environmental transect.* Rept. 83-1. Anthropology Dept., B.P. Bishop Museum, Honolulu. 532 pp.

Wellmon, B.B. 1969. The Parker Ranch, a history. PhD Diss., Texas Christian Univ. Hawaiian Collection, Mookini Library, Univ. Hawaii, Hilo.

Wells, S.M., R.M. Pyle, and N.M. Collins (compilers). 1983. *The IUCN invertebrate red data book.* Internatl. Union Conserv. Nature and Nat. Resour., Gland, Switzerland.

Wester, L. 1983. Vegetataion. pp. 99-114 In J.R. Morgan (ed.), *Hawaii, a geography.* Westview Press, Boulder, Colorado.

Wester, L.L. In press. Origin and distribution of adventive alien flowering plants in Hawai'i. In C.P. Stone, C.W. Smith, and J.T. Tunison (eds.), *Alien plant invasions in native ecosystems of Hawai'i: management and research.* Univ. Hawaii Coop. Natl. Park Resour. Stud. Unit. Univ. Hawaii Press, Honolulu.

Wester, L., and T. Ikagawa. 1988. Weed invasion of *Marsilea villosa* population at 'Ili'ihilauakea Crater, Koko Head, O'ahu. *Newsletter, Hawaiian Bot. Soc.* 27(3):87-101.

Wester, L.L., and H.B. Wood. 1977. Koster's curse *(Clidemia hirta)*, a weed pest in Hawaiian forests. *Environ. Conserv.* 4(1):35-41.

Whiteaker, L.D., and D.E. Gardner. 1985. *The distribution of Myrica faya Ait. in the state of Hawai'i.* Univ. Hawaii Coop. Natl. Park Resour. Stud. Unit Tech. Rept. 55. Botany Dept., Univ. Hawaii, Honolulu. 31 pp.

Whitney, L.D., E.Y. Hosaka, and J.C. Ripperton. 1939 [Reprinted 1964]. *Grasses of the Hawaiian ranges.* Hawaii Agric. Expt. Stn., Univ. Hawaii, Bull. 82. Honolulu. 148 pp.

Whitesell, C.D. 1964. *Silvical characteristics of koa (Acacia koa Gray).* U.S. Forest Service Research Paper PSW-16. Pac. SW Forest and Range Expt. Sta. U.S. Forest Service, Berkeley, Calif. 12 pp.

Wilson, J.T. 1963. A possible origin of the Hawaiian Islands. *Canadian J. Physiol.* 41:863-870.

Wirawan, N. 1974. *Floristic and structural development of native dry forest stands at Mokuleia, northwest Oahu.* US/IBP Univ. Hawaii Tech. Rept. No. 34. Honolulu. 56 pp.

Yamayoshi, H. 1951. Report on hapu [sic] and amaumau fern. Letter to Mr. William Crosby, Territorial Forester, through L.W. Bryan, Associate Forester, March 30. L.W. Bryan Forestry Collection, B.P. Bishop Museum, Honolulu. 10 pp.

Yang, C., D. Murata, and C. Beck (eds.). 1977. Terrestrial plantations. 155 pp. In *Biomass energy for Hawaii*, Vol. IV: *Terrestrial and marine plantations*. Prepared by Hawaii Biomass Energy Study Team, Stanford Univ. and Univ. Hawaii Instit. Energy Stud. Stanford Univ., Stanford, Calif.

Yee, R.S.N., and W.C. Gagné. In press. Activities and needs of the horticulture industry in relation to alien plant problems in Hawai'i. In C.P. Stone, C.W. Smith, and J.T. Tunison (eds.), *Alien plant invasions in native ecosystems of Hawai'i: management and research*. Univ. Hawaii Coop. Natl. Park Resour. Stud. Unit. Univ. Hawaii Press, Honolulu.

Yen, D.E. 1974. *The sweet potato and Oceania.* B.P. Bishop Museum Bull. 236. Bishop Museum Press, Honolulu. 389 pp.

Yen, D.E., P.V. Kirch, P. Rosendahl, and T. Riley. 1972. Prehistoric agriculture in the upper valley of Makaha, Oahu. pp. 39-94 In E.J. Ladd and D.E. Yen (eds.), *Makaha Valley historical project*, Rept. no. 3. Pacific Anthropol. Records 18. Anthropology Dept., B.P. Bishop Museum, Honolulu.

Yoshinaga, A.Y. 1980. *Upper Kipahulu Valley weed survey.* Univ. Hawaii Coop. Natl. Park Resour. Stud. Unit Tech. Rept. 33. Botany Dept., Univ. Hawaii, Honolulu. 17 pp.

Ziegler, M.F.Y. 1989. Kanepu'u: a remnant dry forest on Lana'i, Hawai'i. *'Elepaio* 49(4):19-24.

Zimmerman, E.C. 1941. Argentine ant in Hawaii. *Proc. Hawaiian Entomol. Soc.* 11:108.

Zimmerman, E.C. 1948. *Insects of Hawaii.* Vol. 1. *Introduction.* Univ. Hawaii Press, Honolulu. 206 pp.

Zimmerman, E.C. 1958. *Insects of Hawaii.* Vol. 7. *Macrolepidoptera.* Univ. Hawaii Press, Honolulu.

Zimmerman, E.C. 1978. *Insects of Hawaii.* Vol. 9, *Macrolepidoptera*, Parts I & II. Univ. Hawaii Press, Honolulu. 1903 pp.